T0241402

# SpringerBriefs in Physics

SpringerBriefs in Physics are a series of slim high-quality publications encompassing the entire spectrum of physics. Manuscripts for SpringerBriefs in Physics will be evaluated by Springer and by members of the Editorial Board. Proposals and other communication should be sent to your Publishing Editors at Springer.

Featuring compact volumes of 50 to 125 pages (approximately 20,000–45,000 words), Briefs are shorter than a conventional book but longer than a journal article. Thus, Briefs serve as timely, concise tools for students, researchers, and professionals.

Typical texts for publication might include:

- A snapshot review of the current state of a hot or emerging field
- A concise introduction to core concepts that students must understand in order to make independent contributions
- An extended research report giving more details and discussion than is possible in a conventional journal article
- A manual describing underlying principles and best practices for an experimental technique
- An essay exploring new ideas within physics, related philosophical issues, or broader topics such as science and society

Briefs allow authors to present their ideas and readers to absorb them with minimal time investment. Briefs will be published as part of Springer's eBook collection, with millions of users worldwide. In addition, they will be available, just like other books, for individual print and electronic purchase. Briefs are characterized by fast, global electronic dissemination, straightforward publishing agreements, easy-to-use manuscript preparation and formatting guidelines, and expedited production schedules. We aim for publication 8–12 weeks after acceptance.

Arun Kenath · Chandra Sivaram

# Physics of Gravitational Waves

## Sources and Detection Methods

Arun Kenath
Department of Physics and Electronics
CHRIST (Deemed to be University)
Bengaluru, Karnataka, India

Chandra Sivaram
Indian Institute of Astrophysics
Bengaluru, India

ISSN 2191-5423        ISSN 2191-5431  (electronic)
SpringerBriefs in Physics
ISBN 978-3-031-30462-0        ISBN 978-3-031-30463-7  (eBook)
https://doi.org/10.1007/978-3-031-30463-7

This Springer imprint is published by the registered company Springer Nature Switzerland AG
The registered company address is: Gewerbestrasse 11, 6330 Cham, Switzerland

# Preface

The detection of gravitational waves in 2015 by the Laser Interferometer Gravitational Wave Observatory marked a historic milestone in the field of physics. These ripples in the fabric of space-time, first predicted by Albert Einstein's theory of general relativity hundred years before, have opened up a new way to observe the universe. These cosmic ripples are created by the acceleration of massive objects and travel at the speed of light. They carry information about their sources and give clues about gravity's nature.

Although it took a century for the direct detection of gravitational waves after its prediction, there was indirect evidence for its existence. In 1974, Russell Hulse and Joseph Taylor, Jr. discovered the first binary pulsar, for which they were awarded the Nobel Prize in Physics in 1993. Further observations of the pulsar's timing over the following years revealed a steady decline in the orbital period of this Hulse-Taylor pulsar. This observation was in accordance with the loss of energy and angular momentum predicted by the theory of general relativity through gravitational radiation.

The primary source of the direct detection of gravitational waves so far is from mergers of binary black holes. The release of gravitational waves results in the shrinking of the orbit until the two objects merge into a single, rapidly spinning black hole. This results in a significant release of energy in the form of gravitational waves. Apart from black hole mergers, another important discovery was that of neutron star mergers. Unlike black hole mergers, the merger of neutron stars would also release electromagnetic gamma rays. The simultaneous detection of electromagnetic waves along with gravitational waves has led to the confirmation of some aspects of the fundamental physics involved in the merger of such compact objects. We also discuss novel sources of gravitational waves, like that from dark matter objects (binaries). These could have ramifications to the ongoing dark matter detection experiments.

Another interesting source of gravitational waves is a theoretical type of gravitational wave that is produced by the thermal motion of matter. These thermal gravitational waves are extremely weak and are different from those that current detectors have detected. Thermal gravitational waves are not generally discussed in books on gravitational waves. In this book, we have elaborated on the physics of the production and detection of such waves in detail. The existence, and detection, of these waves, could provide valuable insight into the behaviour of matter at very small scales.

Apart from these aspects, this book also covers areas related to gravitational waves that are not generally discussed. The idea of gravitational waves seems to be intertwined with that of the general theory of relativity. One of the chapters in this book is dedicated to the discussion of gravitational waves in the context of Newtonian physics and the special theory of relativity without invoking the general theory of relativity. Other aspects of gravitational waves that are discussed include possible constraints on dipole radiation predicted by alternate theories of gravity, emissions from pulsars, etc.

We discuss in some detail the various current and future instrumentation to detect gravitational waves of various frequencies. Also possible methods to detect background thermal gravitational radiation are elaborated.

With the continued detection of gravitational waves, a new field of observational astronomy called gravitational wave astronomy was born. Historically, to study the universe, electromagnetic radiation (visible light, X-rays, radio waves, microwaves, etc.) has been exclusively relied upon. Gravitational waves are entirely unrelated to electromagnetic waves; since they interact very weakly with matter (unlike electromagnetic radiation, which can be absorbed, reflected, and refracted), they travel through the universe, carrying information about their origins that is free of any distortions. Some future applications of gravitational wave astronomy, such as the independent estimate of Hubble's constant and constraint on cosmic strings, are elaborated.

Keeping in mind the young research scholars looking to start their careers in gravitational wave physics, we have given a few problems covering all areas related to gravitational waves. Whether you are a student of physics or working in cosmology, this book offers an accessible and engaging introduction to the world of gravitational waves, along with its many interesting facets.

Bengaluru, India                                                                          Arun Kenath
                                                                                        Chandra Sivaram

# Contents

# Chapter 1
# Introduction

**Abstract** In this chapter, we give an overview of the nature and properties of gravitational waves as predicted by the general theory of relativity. Here we highlight the earlier indirect observational evidence for gravitational waves, i.e., from binary pulsars and supermassive black hole binaries like OJ 287.

According to Einstein's general theory of relativity, gravity is how mass deforms the shape of space: near any massive body, the fabric of space becomes curved. But this curving does not always stay near the massive body. In particular, Einstein realised that deformation could propagate throughout the universe, i.e., the gravitational waves transport energy as gravitational radiation. Sources of detectable gravitational waves include binary star systems composed of white dwarfs, neutron stars, or black holes. The existence of gravitational waves is a consequence of the Lorentz invariance of general relativity since it brings the concept of a limiting speed of propagation of the physical interactions with it. Gravitational waves also provide a testing tool for various theories of gravity, i.e. the general relativity and extended theories of gravity, and will provide a definitive test for general relativity.

Although gravitational radiation was directly detected only recently, there was indirect evidence for its existence even earlier. For instance, the 1993 Nobel Prize in physics went for the precision orbital measurements of the Hulse–Taylor binary pulsar system, which provided evidence for the existence of gravitational waves. Various gravitational wave detectors are now in operation; the ongoing LIGO detectors are well into the search, and the space-borne LISA is due to go up within the next two decades. In March 2014, astronomers at the Harvard–Smithsonian Centre for Astrophysics claimed that they had detected and produced "the first direct image of gravitational waves across the primordial sky" within the cosmic microwave background, providing strong evidence for inflation and the Big Bang, but as on January 2015, these results seem to be on shaky grounds.

Gravitational waves (GW) are radiated by objects whose motion involves acceleration, provided that the motion is not perfectly spherically or cylindrically symmetric. Some sources of gravitational waves include two objects orbiting each other (binary pulsars, binary black holes, etc.), a spinning non-axisymmetric object, or a supernova (except in the unlikely event that the explosion is perfectly symmetric). For a source to emit gravitational radiation, the third time derivative of the mass quadrupole moment of an isolated system's stress-energy tensor must be nonzero for it to emit gravitational radiation. This is analogous to the changing dipole moment of charge or current necessary for electromagnetic radiation.

There are many similarities between electrodynamics and gravitation. While electromagnetic waves are oscillations from electromagnetic fields of massive charge, gravitational waves are perturbations in the curvature of space-time. But gravitational waves are different from electromagnetic waves in that the lowest order is the quadrupole (however, the velocity of propagation of gravitational waves is that of the speed of light). The monopole radiation vanishes in both cases because of the conservation of charge and mass (energy), i.e. $\sum e = 0$ (electromagnetic) and $\sum M = 0$ (gravity). For gravity, the dipole term vanishes because of the conservation of momentum. For a gravitational dipole, $d = mr$, the rate of change of momentum, $\ddot{d} = m\ddot{r} = 0$ since the momentum is conserved. So the lowest order is quadrupole radiation. Astronomical GW should begin at $\sim 10^4$ Hz, which is lower than the lowest-frequency astronomical electromagnetic waves. GW will show the details of the bulk motion of dense concentrations of energy, whereas electromagnetic waves will show the thermodynamic state of optically thin concentrations of matter.

The first indirect evidence for GW was first deduced in 1974 for the discovery of binary pulsar system PSR B1913+16 by Hulse and Taylor, and subsequent observations of its energy loss by Taylor and Weisberg demonstrated the existence of GW. The theory says that the energy loss due to GW for this system should decrease its orbital period by $10^{-7}$ s per orbit. Continuing observations of this system showed that the period is decreasing by precisely this amount, thus confirming the existence of GW.

Another indirect test is from the binary black hole system OJ 287, which consists of two black holes of $18 \times 10^9 \, M_\odot$ and $150 \times 10^6 \, M_\odot$. With new observations, astronomers have now characterised the way they whirl about each other in the centre of its galaxy. Its optical light curve has been studied for more than a hundred years, from 1891. Observations have recorded dazzling flares of radiation at semi-regular intervals. The larger black hole is surrounded by a huge accretion disc of dust and gas, which creates radiation, but it is not responsible on its own for the giant flares. The two black holes are on a 12-year orbit, with the smaller black hole not oriented with the plane of the accretion disc; it is on a highly tilted, highly elliptical, processing orbit.

The periodic optical outbursts occur soon after the secondary black hole has impacted the accretion disc of the primary. Since its orbit is irregular, the timing of these flares is different in every orbit. Observation data collected since the late nineteenth century has enabled astronomers to model this orbit and have accurately predicted the most recent two flares. One of these occurred in December 2015, within three weeks of the predicted date. The second flare, nicknamed the Eddington flare, occurred on 31 July 2019. And it was predicted to the day.

Like in the case of binary pulsars, the relativistic gravity effects are considerably larger than in the solar system. The effects of general relativity and strong gravitational field effects would be even more important for binary black holes, such as the binary black hole system in OJ 287, as these are more compact objects. For the orbital period of 12 years and the orbit separation of $\sim 10^4$ AU gives a precession rate of the binary orbit of about forty degrees per period, which is confirmed by observations. The gravitational time delay (the Shapiro effect) for this system is given by, $\frac{4GM}{c^3} \times$ log factor $\sim 10$ days. The energy loss from gravitational waves will cause the orbit to shrink, which for this system is $\sim 0.04$ s for every second of the period. Hence the period between the bursts comes 20 days sooner.

So both the effects observed, i.e. the Shapiro effect and due to gravitational wave energy loss, test the general relativity to within five per cent for such a supermassive binary black hole, 1 Gpc away. The gravitational wave energy released when the merger occurs would be $\sim 10^{55}$ J. Even at Gpc distances, the strain on a gravitational wave detector would be $\sim 10^{-16}$, with a typical frequency of $\sim 5 - 10$ μHz. This would be a very 'bright' signal for future detectors such as the LISA, which has a threshold strain of $\sim 10^{-23}$. We will discuss the physics corresponding to these numbers in later chapters.

# Chapter 2
# Gravitational Waves Sans General Theory of Relativity

**Abstract** The detection of gravitational waves is considered one of the definitive tests of the general theory of relativity. This chapter will examine how Newton's static theory, made special relativistic, can predict gravitational waves. And in the Newtonian framework, we arrive at the quadrupole formula for gravitational waves.

Einstein gave the final version of his general theory of relativity late in 1915. Ten years earlier, in 1905, he had rejuvenated the world of physics with his four landmark papers, including his work on the special theory of relativity. He extended the principle of relativity from frames in uniform relative motion (inertial frames) to frames in accelerated motion. His goal was to ultimately formulate the laws of physics in a form invariant in all arbitrary motion of the frames of reference or described in arbitrary coordinates. This implied that the inertial mass and gravitational mass are identically the same. Gravity is independent of the composition or shape of the test body; the acceleration is the same for all masses.

Soon after setting up the final form of the field equations of general relativity, Einstein predicted the existence of gravitational waves. The linearised field equations (expanded as a perturbation of the flat space-time Lorentz metric) had the form of a relativistic wave equation with the space-time perturbation propagating as a wave (so-called ripples of geometry) moving at the speed of light. The propagating field was a second-order tensor field which in a vacuum has the form, $\Box^2 \phi_{\mu\nu} = 0$, i.e.,

$$\frac{\partial^2 \phi_{\mu\nu}}{\partial t^2} - \frac{1}{c^2}\nabla^2 \phi_{\mu\nu} = 0 \tag{2.1}$$

Newton's gravity theory has a static scalar field $\phi$, so that there is no time dependence of the field; the field obeys, $\nabla^2 \phi = 0$.

However, a wave equation also involves time. So $\phi$ will now be a function of $r$ and $t$, i.e. $\phi = \phi(r, t)$, and the equations would be:

$$\frac{\partial^2 \phi}{\partial t^2} - \frac{1}{c^2}\nabla^2 \phi = 0 \tag{2.2}$$

$c$ being the speed of propagation. Thus if Newton's static theory were made special relativistic, we would naturally have gravitational waves. But these would be scalar. The point here is that the source of the gravitational field to be consistent with special relativity would no longer be the scalar matter density $\rho$, but also include components of the pressure, shear, etc., or in other words, it would be a second-order energy-momentum tensor, $T_{\mu\nu}$. This will have ten components in four-dimensional space-time.

If the source is a second-order tensor, the field for consistency would also be a second-order tensor, i.e. $\phi_{\mu\nu}$ rather than Newtonian $\phi$. To be consistent with special relativity, the field equations of (Newtonian gravity) would be $\Box^2 \phi_{\mu\nu} = 4\pi G T_{\mu\nu}$, rather than Poisson's equation $\nabla^2 \phi = 4\pi G \rho$, which is a static field equation sans waves. As all forms of energy are sources for the gravitational field (and not just the static rest mass density; for example, a moving fluid has all the components), the gravitational field will have ten components (in four dimensions).

For electromagnetic waves, the source is the current four-vector, and we have time-dependent electric and magnetic fields caused by moving electric charges, unlike in the case of the static Coulomb field. Einstein showed Maxwell's equations predicting electromagnetic waves to be Lorentz covariant, i.e. invariant under Lorentz transformations, as any wave equation in space-time should be. After Maxwell's work, it was noticed by some authors like Oliver Heaviside that gravitational field, if made time dependent, should lead to gravitational waves.

Even in the Newtonian framework, one can argue that the lowest order for gravitational waves is the quadrupole, and we can arrive at the quadrupole formula for gravitational waves, albeit with a different numerical coefficient. Like in the case of electromagnetic waves, monopole radiation is ruled out because of mass-energy conservation, i.e. $\sum \dot{e}_i = 0$, $\sum \dot{m}_i = 0$ (dots denote change with time).

For electromagnetism, dipole radiation is given by, $\frac{2}{3c^3}\left|\ddot{D}_e\right|^2$, $D_e = er$, is the dipole moment, charges $e$ separated by $r$. So that $\ddot{D}_e = e\ddot{r}$, so we have the Larmor formula, $P_e = \frac{2e^2}{3c^3}\ddot{r}^2$. Or in terms of frequency (of oscillating charges), $\omega$,

$$P_e = \frac{2e^2}{3c^3}r^2\omega^4 \tag{2.3}$$

Since, $\ddot{r}^2 \approx r^2/t^4 = r^2\omega^4$. I.e. we have a fourth power dependence on the frequency.

For gravitational waves, dipole radiation vanishes because of the conservation of momentum, as seen from, $D_m = mr$, so that $\ddot{D}_m = m\ddot{r}$, but from Newton's second law, this vanishes as $m\ddot{r} = $ force $=$ rate of change of momentum, which vanishes due to momentum conservation. So the lowest order is quadrupole. The mass quadrupole moment is $Q = mr^2$, so that by analogy with electromagnetism (there is as yet no GR here), $P_G = \frac{G}{45c^5}\left|\dddot{Q}\right|^2$. This gives the gravitational radiation power in terms of frequency (for two bodies of mass $m$ orbiting each other) as,

$$P_G = k \frac{G}{c^5} m^2 R^4 \omega^6 \tag{2.4}$$

Apart from the numerical factor (the reader can obtain the factor), this is the same as the quadrupole formula in GR for power emitted in gravitational waves. Other consequences like orbit shrinking and timescale for merger all follow in the same way. All these formulae could have been obtained one or two decades before Einstein's definitive prediction from GR (a bit of hindsight, perhaps).

Gravitational waves produced by accelerated masses of an orbital binary system that propagate as waves outward from their source were first proposed by Oliver Heaviside in 1893 and later by Henri Poincaré in 1905 as waves similar to electromagnetic waves but the gravitational equivalent. Heaviside drew a comparison between the inverse square law of gravity and that of electrostatics, whereas Poincaré put forward the idea of gravitational waves that would originate from a body and travel at the speed of light as a requirement of Lorentz transformations.

It is to be noted that Einstein did not realise the spin 2 nature of the gravitational field. That insight came much later, at least twenty years later, when Pauli and Fierz proposed generalised wave equations for any spin S (as also Bargmann and Wigner). The Pauli–Fierz, spin 2 wave equations are quite identical to the linearised Einstein equations of GR (indeed just what we have given above, with gauge conditions eliminating the spin 1 and spin 0 fields, i.e. $\partial^\nu \phi_{\mu\nu} = 0$, $\phi_\mu^\mu = 0$). The linearised equations also satisfy the transverse traceless TT conditions or gauges, like in the case of electromagnetic waves.

# Chapter 3
# Gravitational Waves from Binary Systems

**Abstract** We discuss the emission of gravitational waves from binary systems of compact objects and associated phenomena such as the shrinking of the orbits and the merger times. Gravitational radiation associated with supermassive black holes is also covered. Apart from the usual binary systems, we also consider new sources like binary dark matter objects. The expected gravitational wave backgrounds from these various sources (at different frequencies) are estimated.

Gravitational waves carry energy away from their sources, and in the case of orbiting bodies, this is associated with an inspiral or decrease in orbit. The energy loss due to gravitational radiation, for a system of two equal masses $M$ separated by a distance $a$, is given as,

$$\dot{E}_{GW} = \frac{32G}{5c^5} M^2 a^4 \omega^6 f(e) \tag{3.1}$$

where $f(e)$ is a function of orbital eccentricity and $\omega$ is the angular frequency. For a system like the Sun and the Earth, the power is about 200 W. This is extremely small compared to the total electromagnetic radiation given off by the Sun ($\sim 4 \times 10^{26}$ W).

Gravitational radiation carries away energy, and as a consequence, the radius of the orbit gradually shrinks. As the distance between the bodies decreases, they revolve more rapidly in their orbit. The overall angular momentum is reduced, and this reduction corresponds to the angular momentum carried off by gravitational radiation. As the radius decreases, the power lost to gravitational radiation increases even more, as seen from the earlier equation. For the Sun and Earth system, the Earth's orbit shrinks by $10^{-20}$ m/s, which is about $3.5 \times 10^{-13}$ m per year. The effect of gravitational radiation on the Earth's orbit is negligible, but that is not the case for larger objects with closer orbits. In general, we have,

$$\left( T_{\mu\nu} \right)_{GW} = \frac{c^4}{32\pi G} \left\langle \overline{h}_{\alpha\beta,\mu} \overline{h}^{\alpha\beta,\nu} - \frac{1}{2} h_{,\mu} h^{,\mu} \right\rangle \tag{3.2}$$

Here, $h_{\mu\nu}$ is the perturbation from Minkowski space ($g_{\mu\nu} = \eta_{\mu\nu} + h_{\mu\nu}$). In the Transverse Traceless gauge, $\frac{1}{2}h_{,\mu}h^{,\mu} = 0$. And we have,

$$\dot{E}_{GW} = -\frac{G}{45c^5}\langle \dddot{Q}_{ij}\dddot{Q}_{ij}\rangle \tag{3.3}$$

$\dddot{Q}_{ij}$ is the third time derivative of the quadrupole moment tensor averaged over a period of an oscillation. The strain at a distance $r$ is $h \sim \frac{G}{c^4}\frac{\ddot{Q}}{r}$. For a rod of length $L$, mass $M$, spinning at a frequency $\omega$,

$$P_{GW} = \frac{2G}{45c^5}M^2 L^4 \omega^6 \approx 10^{-54}\,\mathrm{W}\left(\frac{M}{1\,\mathrm{kg}}\right)^2\left(\frac{L}{1\,\mathrm{m}}\right)^4\left(\frac{\omega}{1\,\mathrm{Hz}}\right)^6 \tag{3.4}$$

For a binary of mass $M$ or compact collapsing object of radius $R$, characteristic frequency is, $\omega^2 = \frac{GM}{R^3}$ and $P_{GW} \approx \frac{c^5}{G}\left(\frac{GM}{Rc^2}\right)^5$. The maximum power due to gravitational waves is,

$$(P_{GW})_{\max} \leq \frac{c^5}{G} \approx 10^{52.5}\,\mathrm{W} \tag{3.5}$$

Most of the energy is radiated in the last orbital period, $\Delta E \sim \epsilon M c^2$, where $\epsilon \sim 10^{-3}$. The characteristic frequency of gravitational waves emitted by a very compact object with $\frac{GM}{Rc^2} \sim 1$, is,

$$\omega = \frac{c^3}{GM} \approx \left(\frac{M}{10^5 M_\odot}\right)^{-1}\,\mathrm{Hz} \tag{3.6}$$

For a solar mass object, it would be $\sim 100\,\mathrm{kHz}$.

If two objects (binary system) with total mass $M$ and reduced mass $\mu$, are separated initially by a distance $a_i$, then the orbital decay time (merger time) is,

$$\tau_{merger} = \frac{5c^5}{256G^3}\frac{a_i^4}{\mu M^2} \tag{3.7}$$

For the formation of a neutron star in a Type II supernova (collapse of a massive star), we have $\frac{GM}{Rc^2} \sim 0.1$, and one can expect kHz frequency bursts of gravitational waves with $P_{GW} \leq 10^{-5}\left(\frac{c^5}{G}\right)$ and strain of $h \sim \frac{GE}{rc^4}$, with $r$ being the distance to the source.

## 3.1 Gravitational Waves from Binary Black Holes

About half of all stars in the universe come in binary systems. If they are sufficiently massive to start with, then once their nuclear reaction terminates, they end up as binary neutron stars or black holes. This binary system emits gravitational radiation, which carries away energy, as given by Eq. (3.1). As a consequence, the orbit shrinks, slowly at first but faster and faster, until the two objects merge into a single, rapidly spinning black hole. This process releases a tremendous amount of energy in the form of gravitational waves. These waves travel outward in all directions at the speed of light, carrying information about the black holes and the merger event with them.

We can analyse the system using classical laws such as Kepler's equations. The equation of force for the stabilized system is given by equating the centrifugal force to the gravitational force. From Kepler's third law, the period, $P$, for a binary system of masses $M_1$ and $M_2$, is related to the separation, $a$, by,

$$P^2 = \frac{4\pi^2}{G(M_1 + M_2)} a^3 \tag{3.8}$$

For a system of binary black holes of mass $10^8\, M_\odot$ each, separated by one parsec, the orbital period is $\sim 10^3$ years. Knowing the period, we can determine the orbital velocity $v$, from $vP = 2\pi R$. For this system, this works out to $v \approx 10^5\,\mathrm{m/s}$. The energy lost by the system due to the emission of gravitational waves (assuming a circular orbit) is,

$$\dot{E}_{GW} = \frac{128 v^{10}}{5 G C^5} \tag{3.9}$$

For the above system, this works out to $\dot{E}_{GW} \approx 10^{25}\,\mathrm{J/s}$. The system emitting gravitational waves at this rate will merge together in a time scale given by,

$$\tau_{merger} = \frac{5}{64} \frac{c^5}{G^3} \frac{a^4}{M^2 \mu} \tag{3.10}$$

Here $M = M_1 + M_2$ and $\mu$ is the reduced mass. This works out to be $\tau_{merger} \approx 10^6\, T_H$ for the above-mentioned system, where $T_H = 10^{10}$ years is the Hubble time. The total energy emitted by the system during the merger is $\dot{E}_{GW} \times \tau_{merger} \approx 10^{50}$ J.

This is the amount of energy released during the process of the merger of two black holes. Most of the energy liberated is in the last stage of the merger when the black holes are almost in contact. At this stage, the distance of separation between the two black holes corresponds to twice the Schwarzschild radius of the black hole, that is,

$$a = 2\left(\frac{2GM}{c^2}\right) \tag{3.11}$$

And the energy emitted during this stage is,

$$E = \frac{GM^2}{a} = \frac{1}{4}Mc^2 \tag{3.12}$$

And this corresponds to about $10^{50}$ J.

For the merger time to be of the order of the Hubble time, i.e. $t_{merger} = t_H \left( \approx 4.35 \times 10^{17} \text{ s} \right)$, we have the following Eq. (3.13), $t_H M^3 = constant \times a^4$. For instance, in an equal mass binary, each with 2 $M_\odot$, would merge in Hubble time if their initial separation is $\approx 2.5 \times 10^9$ m $\left( = 2.5 \times 10^{11} \text{ cm} \right)$. The maximum initial separation $(a_i)$ of the equal mass binary of different mass range to merge within the Hubble time is given in Fig. 3.1.

From this, we can express the separation in terms of the mass as:

$$a = \left( \frac{t_H M^3}{constant} \right)^{1/4} = constant' \times M^{3/4} \tag{3.13}$$

From Eqs. (3.6) and (3.13), we have the frequency of GW in terms of the mass $M$ as:

$$\omega = constant'' \times M^{-5/8} \tag{3.14}$$

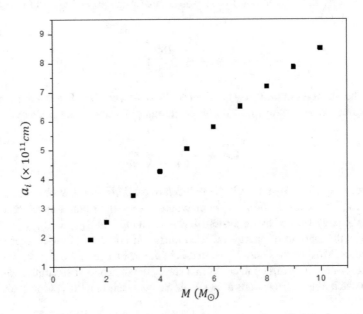

**Fig. 3.1** Initial separation for various masses for $t_{merger} = t_H$ (image credit: Journal of High Energy Physics, Gravitation and Cosmology, 7, 698, 2021)

With the constants, this scaling of $\omega$ with mass $M$ is given as:

$$\omega = 2.7 \times 10^{-4}\,\text{Hz} \left( \frac{M}{M_\odot} \right)^{-5/8} \tag{3.15}$$

From Eqs. (3.13) and (3.14), we have $a^4 \propto M^3$ and $\omega^6 \propto M^{-15/4}$ respectively. These, along with Eq. (3.4), give the mass dependence of power, $P$ as, $P \propto M^{5/4}$. With the constants, this scaling is given as:

$$P = 1.5 \times 10^{24}\,\text{J/s} \left( \frac{M}{M_\odot} \right)^{5/4} \tag{3.16}$$

Figure 3.2a gives the variation in frequency ($\omega$) of GW with mass ($M$) and Fig. 3.2b gives the variation of power ($P$) emitted in GW with mass ($M$). The total energy emitted over the Hubble time ($t_H$) is $E_H = t_H \times P$. This will give a background flux which could affect gravitational waves measured from distant bursts. These values for varying masses of the equal-mass binary black holes are tabulated in Table 3.1. These values of frequency and power are those associated with the initial stages of the separation of the equal mass binaries. They evolve over time as the separation decreases, with both $\omega$ and $P$ increasing with decreasing separation. The frequency detected by gravitational wave detectors, like LIGO, is that during the merger as the separation, $a \approx 2 \times$ radius of the neutron star (NS), or $2 \times$ Schwarzschild radius in the case of black holes (BH).

As the binary components continue to approach each other, the frequency increases, culminating with the chirp frequency as they eventually merge, which is the one measured by GW detectors. The rate of change of the frequency with decreasing distance between the binary components is given by:

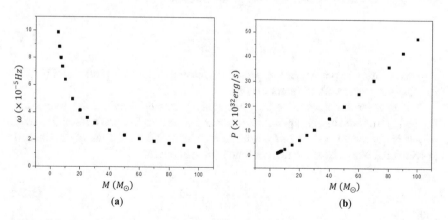

**Fig. 3.2** Variation with the mass of the binary components in the: **a** frequency of GW; **b** power emitted in GW (image credit: Journal of High Energy Physics, Gravitation and Cosmology, 7, 698, 2021)

**Table 3.1** Variation with BH mass of frequency, power, and total energy over Hubble time emitted in GW

| $M\ (M_\odot)$ | $\omega$ (Hz) | $P$ (J/s) | $E_H$ (J) |
| --- | --- | --- | --- |
| 5 | $9.87 \times 10^{-5}$ | $1.12 \times 10^{25}$ | $4.87 \times 10^{42}$ |
| 6 | $8.81 \times 10^{-5}$ | $1.41 \times 10^{25}$ | $6.13 \times 10^{42}$ |
| 7 | $8.00 \times 10^{-5}$ | $1.71 \times 10^{25}$ | $7.44 \times 10^{42}$ |
| 8 | $7.36 \times 10^{-5}$ | $2.02 \times 10^{25}$ | $8.79 \times 10^{42}$ |
| 10 | $6.40 \times 10^{-5}$ | $2.67 \times 10^{25}$ | $1.16 \times 10^{43}$ |
| 15 | $4.97 \times 10^{-5}$ | $4.43 \times 10^{25}$ | $1.93 \times 10^{43}$ |
| 20 | $4.15 \times 10^{-5}$ | $6.34 \times 10^{25}$ | $2.76 \times 10^{43}$ |
| 25 | $3.61 \times 10^{-5}$ | $8.38 \times 10^{25}$ | $3.64 \times 10^{43}$ |
| 30 | $3.22 \times 10^{-5}$ | $1.05 \times 10^{26}$ | $4.57 \times 10^{43}$ |
| 40 | $2.69 \times 10^{-5}$ | $1.51 \times 10^{26}$ | $6.57 \times 10^{43}$ |
| 50 | $2.34 \times 10^{-5}$ | $1.99 \times 10^{26}$ | $8.66 \times 10^{43}$ |
| 60 | $2.09 \times 10^{-5}$ | $2.50 \times 10^{26}$ | $1.09 \times 10^{44}$ |
| 70 | $1.90 \times 10^{-5}$ | $3.04 \times 10^{26}$ | $1.32 \times 10^{44}$ |
| 80 | $1.75 \times 10^{-5}$ | $3.59 \times 10^{26}$ | $1.56 \times 10^{44}$ |
| 90 | $1.62 \times 10^{-5}$ | $4.16 \times 10^{26}$ | $1.81 \times 10^{44}$ |
| 100 | $1.52 \times 10^{-5}$ | $4.74 \times 10^{26}$ | $2.06 \times 10^{44}$ |

$$\frac{d\omega}{da} = \left(\frac{9GM}{2a^5}\right)^{1/2} \tag{3.17}$$

The variation in frequency—for binary systems with different masses—with respect to decreasing distance (as they approach) is plotted in Fig. 3.3a. As the binary components continue to inspiral, the power emitted due to gravitational radiation increases, and the corresponding change in power is:

$$\frac{dP}{da} = \frac{32G^4M^5}{c^5 a^6} \tag{3.18}$$

The variation in power with respect to decreasing distance is plotted in Fig. 3.3b for binary systems with different masses.

The total power radiated during the merger can be obtained by integrating Eq. (3.18) from the initial separation to that at the merger ($2\times$ Schwarzschild radius of the black holes). For a $2\,M_\odot$ neutron star binary, the total energy emitted in gravitational waves over the merger time is then given as:

$$E = \frac{2.7 \times 10^{47}}{t\ (\text{in s})^{1/4}}\ \text{J} \tag{3.19}$$

For a merger time of about a billion years, the total energy emitted is $E \approx 10^{44}$ J. With one NS merger per galaxy in $10^5$ years, the total number of mergers in the whole

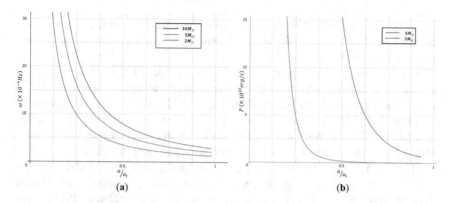

**Fig. 3.3** Variation with decreasing distance between equal mass binaries (of different masses) in the: **a** frequency; **b** power radiated (image credit: Journal of High Energy Physics, Gravitation and Cosmology, 7, 698, 2021)

universe (with $\sim 10^{11}$ galaxies) is $\sim 10^6$ per year. Then in the Hubble time, the total mergers that have already taken place would be $\sim 10^{16}$. Each of these mergers would release an energy of $\sim 10^{45}$ J ($= GM^2/R$, the binding energy of the NS, where $R$ is the radius of the NS) in GW. This then implies that total energy of $\sim 10^{61}$ J is released in GW. This is spread over the Hubble volume of $2\pi^2 R_H^3 \approx 10^{79}$ m$^3$, hence the energy density is $\sim 10^{-18}$ J/m$^3$ and the corresponding flux on Earth is $\sim 10^{-12}$ J/m$^2$/s in the frequency range of 100–1000 Hz (the chirp frequency detected by the GW detectors at the final stages of the merger). The flux of gravitational waves is related to the strain as:

$$ f = \frac{c^3}{32\pi G}\omega^2 h^2 \tag{3.20} $$

For a frequency of 100 Hz, the strain will be: $h \approx 10^{-24}$. This is the strain due to the background flux, and it is about one order lesser than the sensitivity of LIGO, which is $\approx 10^{-23}$.

In the case of BH, a 20–30 $M_\odot$ black holes are formed by the collapse of a 50 $M_\odot$ progenitor star. On average, a massive star collapses every second in the universe. So in the Hubble time, a total of $\sim 10^{17}$ such collapses will occur. The number of stars from 20 to 50 $M_\odot$ mass range is given by (using the usual Salpeter mass function):

$$ N = \int_{20\,M_\odot}^{50\,M_\odot} \frac{k\,dM}{M^{2.5}} \tag{3.21} $$

where $k \sim 1$ star/$(\text{pc}^3\,M_\odot)$, this works out to be $N = 0.02$ star/$(\text{pc}^3\,M_\odot)$.

Stellar mass BH mergers are estimated to occur once every 200–300 s in the universe. So the total number of such mergers in the Hubble time will be $\sim 10^{15}$.

The energy flux is comparable to NS mergers. The energy released is 10–100 times more than that in the case of NS mergers, but the frequency is 100 times lower; hence the background strain is comparable.

In the case of mergers of super massive black holes (SMBH), the frequency is very low, but the energies are much higher, so only LISA can detect the corresponding background strain. For example, the binary system OJ 287 consists of two orbiting supermassive black holes about a Gpc away. The mass of the primary black hole is deduced to be 18 billion solar mass, and the secondary black hole of OJ 287 has a mass of 150 million solar mass. For the orbital period of 12 years, the orbit separation is $\sim 10^4$ AU. The energy released as gravitational radiation when the merger occurs would be $\sim 10^{55}$ J. Even at Gpc distance, the strain on a gravitational wave detector would be, $h \approx 10^{-16}$, with a typical frequency of $\sim$ 5–10 $\mu$Hz. This would be a very 'bright' signal for LISA (which has a threshold of $\sim 10^{-23}$).

The first gravitational wave detection by LIGO was published on 11 February 2016. It was the first direct detection of gravitational waves and provided evidence for the observation of a binary black hole merger. On September 14, 2015, two detectors of LIGO observed a transient gravitational-wave signal of frequency from 35 to 250 Hz with a peak $h \sim 10^{-21}$ of a two black hole merger. The initial black hole masses were about 36 $M_\odot$ and 29 $M_\odot$, with the merger creating a final black hole of mass of $\sim$ 62 $M_\odot$. The rest, $\sim$ 3 $M_\odot$, was radiated as gravitational waves. GW150914 arrived first at L1 (at Livingston) and about 6.9 ms later at H1 (at Hanford).

Ever since this first discovery, gravitational waves have been regularly detected mostly from the merger of massive black holes in binary systems. Among the dozens of events, only two have been attributed to the merger of neutron stars. Many of the events from pairs of colliding black holes are attributed to arise from black holes with several tens of solar masses, i.e. around 30–50 $M_\odot$, including the earliest events. A puzzling aspect is that the inspiralling dance of a pair of such massive black holes should last for billions of years (longer than the Hubble age, from Eq. (3.10)), but yet we have 'caught' about a few dozen such collisions in over six years.

A well-known binary system in our galaxy is Plaskett's Star, consisting of a pair of massive stars in that mass range which could collapse into black holes at the end point of their evolution. The two stars (current measurements imply both stars of the binary are blue supergiants) have a mass of around 50 $M_\odot$ each, and orbiting with a period of about 14 days with an orbital velocity of 220 km/s. This implies a separation between them of about 0.5 AU ($\sim 6 \times 10^{10}$ m). Each of these stars would end up as a black hole of around 30–40 $M_\odot$. Assuming that the separation is not much changed (indeed, if they become red supergiants, their separation should increase due to mass loss etc.), i.e. around 0.5–1.0 AU, the merger time due to gravitational radiation when they end up as massive stellar black hole (from Eq. (3.10)), is $\tau_{merger} \approx 3 \times 10^{18}$ s, about ten times the Hubble age. Even if the orbits are highly eccentric, $e = 0.5$, this would still be three times the Hubble age.

Eta Carinae is another example, which is also a binary. The well-known primary star is perhaps the most massive star in our galaxy, with a mass of 120 $M_\odot$. A companion was discovered some years back with a mass of 30 $M_\odot$ and an orbital period of 5 years. Such a system is another example where a pair of massive stellar binary black holes will form (in a few million years). For the separation corresponding to the above orbital period, this black hole pair (from Eq. (3.10)) implies a merger time of $10^{20}$ years. Considering the above time scales for the massive stellar progenitors (potential candidates), it is a bit surprising that several such massive black hole binary mergers have been detected in the last few years. So some processes must be at work to accelerate the collision process to make the black holes come together more quickly than anticipated.

When the progenitor stars evolved into red supergiants (like Plaskett's Star) and orbiting close, one could be subsumed into the other, and the pair will collide before ever becoming black holes. This implies that any pair of large black holes must start their existence much farther apart, indeed so far apart that collisions and mergers (leading to gravitational waves) would be extremely rare (as seen above, the merger times, even for eccentric orbits, are much larger than Hubble age). Yet such mergers (collisions) seem to be fairly common, considering so many such events seen. How can we set tight limits on black hole pairs that were never tight-knit supergiant stellar pairs? Can some massive stars collapse into black holes without ballooning into red supergiants (like in the Humphrey–Davidson effect), or can solitary massive stellar black holes meet and bind to turn pairs? How frequent is the latter process? Is there something missing in our understanding of the evolution of very massive stars?

Another idea suggests that under the right conditions, a third object can trigger a process that brings in a pair of objects closer together. Three body effects provide a way for far-apart massive stars to first collapse into black holes, which then draw close enough to collide. Many massive stars are in triple systems, so perhaps such 3-body effects must be considered. But then, can all the events observed be due to such effects? It is unclear how often orbits in triple star systems will be 'large-angled' enough to trigger the effect.

For instance, say a third object rotates around the Earth-Moon system at an angle so that orbits are not aligned. If the angle between orbits is large enough, gravitation effects from the third object could interact with the orbits of Earth and Moon. Their paths would stretch out into long ellipses taking objects farther apart before swinging them much closer and eventually leading to collapse. For black holes, the third object could be a stellar-mass black hole or a massive star yet to collapse. It could even be one of the super massive black holes at the galactic centre. In this situation, two massive stars in the galactic centre collapse to form black holes. The pair of black holes plus the SMBH make a three-body system. However, all these proposals have many loopholes and missing features and processes to be figured out.

On 21 May 2019, interferometric instruments in both the LIGO and VIRGO gravitational wave detectors were all triggered by a sharp signal lasting just one-tenth of a second (0.1 s). It marked the final stage moments of the merger of two inspiralling black holes, one with 66 $M_\odot$ and the other with 85 $M_\odot$. The energy equivalent of 8 solar mass was released in an instant (0.1 s), corresponding to a power output of around $10^{49}$ W. This was the biggest and most energetic merger yet observed of binary black holes. The signal travelled for seven billion light years to reach Earth and was still strong enough to rattle the laser detectors of LIGO (in the US) and VIRGO (in Italy). The two merging black holes produced a single black hole with a mass of 142 $M_\odot$. It could signal the existence of a new class of Intermediate Mass Black Holes (IMBH) in the range of 100–1000 solar masses.

The event, dubbed GW 190521, occurred when the universe was around half its current age of 13.8 billion years and is one of the latest to come out of the LIGO-VIRGO collaborations, which so far has registered and examined over fifty GW events. The increased sensitivity of the detectors could result in more than one event being found every day.

As far as this event is concerned, the involvement of an 85 $M_\odot$ black hole is not quite consistent with our understanding of how such massive objects form. The physical process operating in very massive stars suggests that the formation of black holes in the mass range between 65 $M_\odot$ and 120 $M_\odot$ is not possible. The dying progenitor massive stars that could yield such black holes are actually expected to tear or blow themselves apart, leaving no remnant behind (like in a Type Ia SN, where nuclear detonation of a C–O white dwarf core blows the object into smithereens). In particular, models identify a range of masses between 65 and 130 $M_\odot$, called the pair instantly mass gap in which it is thought that black holes cannot form by collapsing stars since more nuclear energy is released than the star's gravitational binding energy, tearing it apart. The massive star is completely disrupted, and no remnant (like a black hole) is left behind. Core collapse supernovae (SN) can explain how stars as massive as 130 $M_\odot$ can produce black holes of up to 65 $M_\odot$.

For heavier stars, a phenomenon known as 'pair instability' is thought to occur when the core photons become so energetic that they can transform into electron–positron pairs ($e^+e^-$ pairs). The formation of these pairs lowers the radiation pressure exerted by the photons causing the star to become unstable (to gravitational collapse) against collapse, resulting in an explosive release of titanic energy strong enough to disrupt the star, leaving no remnant behind. Even more massive stars, greater than 200 $M_\odot$ would eventually collapse into a black hole of 120 $M_\odot$ directly. The theoretical implications are that collapsing stars should not be able to produce black holes between 65 and 120 $M_\odot$, a range known as the pair-instability mass gap. Now the heavier of the two black holes in the latest GW observation, i.e. at 85 $M_\odot$ is the first detected so far within the pair-instability gap. So how did these two black holes with such masses originate? It is conjectured that they themselves could have formed by earlier mergers of two smaller mass black holes. This raises the question of a hierarchy of mergers as possible paths to form higher and bigger black holes. So this final 142 $M_\odot$ black hole forming in this event may go on to merge with other very massive black holes. Can the build-up process go all the way to form SMBH?

Recent studies have pointed out that stellar structure and evolution could be affected if there is an admixture of dark matter (DM) particles with baryonic (gaseous) matter, especially in the early universe. This has also been recently extended to study the evolution of low-mass red giants. Even a small fraction of DM in the stellar material will end up raising core temperatures, speeding up the evolution. Here we suggest that the admixture of DM could also affect the later nuclear-burning stages of very massive stars.

For instance, at the C burning stage, when the temperature of the core has crossed 700 million degrees, the photoneutrino process becomes important via $\gamma + e^- \rightarrow$

$e^- + \nu_e + \bar{\nu}_e$, photons going into neutrino pairs. These neutrinos drain away energy from the star. This process releases $10^2$ J/kg, while the C burning provides $6 \times 10^{13}$ J/kg. The photo neutrino process goes as the 8th power of temperature, i.e. as $T_C^8$ ($T_C$ is the core temperature). If $T_C$ is say doubled due to the presence of dark matter (as basically DM particles don't get heated but add to gravity), the drainage of energy can increase 300 times. So the star could collapse even before fusing heavier nuclei (e.g. O or Mg burning) when temperatures reach $3 \times 10^{9\circ}$ for pair-instability to set in. Moreover $\nu$ pair production at these temperatures (going as $T_C^9$) could rapidly drain away the energy from the core, making the star collapse rather than explode.

Another gravitational wave detection of note was in 2020 when the LIGO Scientific Collaboration and Virgo Collaboration reported the observation of gravitational waves from a binary black hole coalescence labelled GW190412. This binary is different from the earlier observations due to its asymmetric masses, here a 30 $M_\odot$ black hole merged with an 8 $M_\odot$ companion. The heavier black hole rotated with a dimensionless spin magnitude between 0.17 and 0.59. Asymmetric systems are predicted to emit gravitational waves with stronger contributions from higher multipoles, which was seen in the observed signal, with strong evidence for gravitational radiation beyond the leading quadrupolar order.

This asymmetry in their mass made the larger black hole distort the space around it, so the smaller black hole's trajectory deviated from a perfect spiral. All the other mergers, with nearly equal masses, produced a wave that forms a similar 'chirp' shape which increases in intensity and frequency, leading to the moment of collision. But the signal from GW190412 was different in the sense that its intensity didn't simply rise as in a chirp. Such type of events is exciting since they provide new, more precise ways of testing the general theory of relativity. In particular, this data enables us to discern the 'spin' of a black hole. Observations of black holes in the Milky Way suggested that black holes should have high spins. But the previous events from the first two runs of the LIGO detector did not show any indication of this in the gravitational wave data. Detecting spins can shed light on how the black holes formed and came to orbit each other.

## 3.2 Gravitational Waves from Compact Binary Stars

Neutron stars are the incredibly dense cores of collapsed stars and are composed almost entirely of neutrons. They are incredibly massive, with a single neutron star having about the same mass as the Sun but only about the size of a small city. When two of these incredibly dense objects collide, they create a massive explosion known as a kilonova, which can outshine an entire galaxy for a brief period of time. The event also produces gravitational waves, which can be detected by instruments such as LIGO.

Another puzzling aspect of the hitherto detected gravitation wave events is the paucity of neutron star mergers. Their merger times would be shorter, as in the case of the 2.4-h binary pulsar. Neutron stars are incredibly dense objects (when the star is crushed when it collapses gravitationally, the protons in the nucleus and the electrons are squeezed together, resulting in most of the matter becoming all neutrons). A teaspoon of such an object would weigh a billion tons. When neutron stars merge, the final coalescence takes place over a period of a millisecond releasing a billion trillion trillion trillion joules of energy. However, unlike black hole mergers, the merger of neutron stars would also lead to a tremendous explosive release of electromagnetic gamma-rays (a so-called short-duration gamma-ray burst lasting a second or two).

This two neutron star binary with individual masses of $1.2\,M_\odot$ and a separation of twice the Earth-Moon distance would have a merger time of around $10^8$ years. Neutron star merger estimates imply $10^{-4}$ mergers per galaxy per year, i.e. one merger every year in 10,000 galaxies. So considering thousands of galaxies in the Virgo Cluster, 15 Mpc away, we should have seen several events in the past six years. Although neutron star binary systems are expected to be more numerous than black hole binary systems, the energy release of gravitational waves may be too weak to be detected (by LIGO) unless they are much nearer. Indeed the massive black hole merger events all took place at some billion or more light years away, whereas the neutron star merger, detected in 2019, took place in the galaxy NGC 4993, which is about ten times nearer at only 130 million light years away. In fact, one such merger is expected in a cluster of ten thousand galaxies (like the Virgo cluster) every year. As really massive stars leading to massive black holes are very rare, there should be many more neutron star binaries. However, the merger of massive black holes releases much more energy (as gravitational waves) so that they can be detected from much greater distances.

The neutron star merger of August 2017 was the most spectacular one, accompanied by a gamma-ray burst, whose optical afterglow revealed spectra of heavy elements like gold, and platinum, i.e. r-process elements produced in the merger of two neutron stars. That event was 45 Mpc away. The more recent merger, also involving a neutron star binary, was not associated with an electromagnetic signal. The binary had an estimated mass of $3.5\,M_\odot$, much heavier than any NS binary in our galaxy. Perhaps the merger led to a black hole 'drowning out' any electromagnetic signal. Furthermore, the associated neutrino signal in neutron star mergers would be too weak to be detected even at Virgo Cluster distance. A high energy neutrino event was associated with a black hole merger but was not confirmed. Ideally, we would like to observe simultaneously gravitational wave, electromagnetic, and neutrino signals from NS mergers. So far, only one event has led to the simultaneous detection of GW and electromagnetic signals.

Such neutron star mergers were also expected to produce the heaviest elements in the periodic table, like gold, platinum, and uranium. Now observations of the electromagnetic radiation from the merger have revealed tell-tale scattering from these heavy rare chemical elements. Nuclei of gold, platinum, etc., are being ejected into space at high speeds. Thermonuclear reactions that power stars in their lifetime do not produce elements heavier than iron (stars like the Sun do not even produce elements beyond carbon and oxygen). The heavier than iron elements are produced by successive neutron capture (as a large number of neutrons are released in neutron star formation or merger) on the iron group elements. These successive neutron captures and subsequent beta decay keep increasing the atomic and mass numbers of the nuclei. As the neutron density is so large, the reactions take place on very short time scales. With the simultaneous release of gravitational waves and gamma-rays from the neutron star mergers, all these theories are now nicely confirmed, including the model for short-duration gamma-ray bursts. This confirms long-debated mysteries on the origin of the heaviest rare elements like gold, platinum, and uranium.

Another discovery in 2020 was that of two white dwarfs spiralling around each other, and they might be producing gravitational waves. The study of this system will not only advance our understanding of white dwarf binaries and gravitational wave sources, but it will also be important in validating the efficiency of the LISA (Laser Interferometer Space Antenna) gravitational wave observatory, that is expected to be launch in the 2030s. White dwarfs are the final stage in the evolution of a star like the Sun and are characterized by their small size and high density. White dwarfs are made up mostly of carbon and oxygen, with an initial surface temperature of around 100,000 °C. They are extremely dense, with a mass similar to that of the Sun but a size comparable to that of the Earth.

This system, known as J2322+0509, has a short orbital period of 1201 s and is the first gravitational wave source of its kind ever identified. It is the first time a binary white dwarf system made up of two helium core white dwarf stars that are clearly separated is detected. Since this binary had no light curve, the system was tricky to spot, but it was an extremely strong source of gravitational waves. The gravitational waves that are being emitted are causing the pair to lose energy; in six or seven million years, they will merge into a single, more massive white dwarf.

## 3.3 Gravitational Waves from Dark Matter Objects

Dark matter is a hypothetical form of matter that is thought to make up approximately 85% of the matter in the universe. It is *dark* because it does not interact with any forms of electromagnetic radiation, making it difficult to detect. Despite its elusive nature, dark matter is thought to play a crucial role in the way galaxies and galaxy clusters form and evolve. The evidence for the existence of such non-radiating matter goes back more than eighty years ago when Zwicky (in 1937) was trying to estimate the masses of large clusters of galaxies. Surprisingly it was found that the dynamical mass of the cluster, deduced from the motion of the galaxies (i.e. their dispersion of velocities), in a large cluster of galaxies was at least a hundred times their luminous mass. This led Zwicky to conclude that most of the matter in such clusters is not made up of luminous objects like stars or clusters of stars but consists of matter which does not radiate.

Later observations starting about forty years ago and continuing till now also revealed unmistakably that even individual galaxies like our Milky Way are dominated by DM. We know this for spiral galaxies because it turns out that objects orbiting these galaxies at larger distances from the galactic centre move around more or less the same velocity as objects much closer to the centre, contrary to what is expected. This suggests that the mass in the galaxy is still growing even after light dies out.

Apart from the velocity distribution of galaxies and galaxy clusters, there are other pieces of evidence pointing to the presence of dark matter. Extended emission in X-ray observations of clusters of galaxies indicates the presence of hot gas distributed throughout the cluster volume. Results from Chandra X-ray Observatory on the distribution of dark matter in a massive cluster of galaxies (such as Abell 2029, which consists of thousands of galaxies surrounded by a huge cloud of hot gas) indicate that the cluster is primarily held together by the gravity of the dark matter.

Another method to detect the presence of DM is gravitational lensing. This method provides an alternate method of measuring the mass of the cluster without relying on observations of the dynamics of the cluster. All these different methods point to the currently accepted scenario that ~ 80% of the total amount of matter in the galaxy is a form of DM. There is no shortage of ideas as to what dark matter could be. Serious candidates have been proposed with masses ranging from $10^{-5}$ eV ($10^{-71}$ $M_\odot$) particles to $10^4$ $M_\odot$ black holes. That's a range of masses of over 75 orders of magnitude.

Baryonic DM candidates are cosmologically insignificant; hence much of the focus is primarily on non-baryonic candidates. The non-baryonic candidates are basically elementary particles which are either not yet discovered or have non-standard properties. There is compelling evidence that much of the DM may be made up of as yet undiscovered particles like axions, neutralinos, gravitinos or composites of the same. The current observations of large scale structures indicate that the DM is 'cold' (CDM). These particles are those described by a non-relativistic equation of state at the time when galaxies could just start to form. Some weakly interacting massive particles (WIMPs) could be CDM.

These DM particles, each of several GeV rest mass, could gravitationally condense and form degenerate objects of planetary mass. These objects made up mostly of DM particles of mass, $m_D$, will have a typical mass given as,

$$M = \frac{M_{Pl}^3}{m_D^2} \tag{3.22}$$

where, $M_{Pl} = \left(\frac{\hbar c}{G}\right)^{1/2} \approx 2 \times 10^{-8}$ kg is the Planck mass.

There is some interest in the detection of excess gamma-rays from the galactic centre, which is attributed to the decay of 60 GeV DM particles. This mass for the DM particle is also favoured from other results (like the DAMA experiment, among others). Thus if we put $m_D = 60$ GeV in Eq. (3.22), the mass of these objects is,

$$M \approx 10^{26} \text{ kg} \tag{3.23}$$

which is about the Neptune mass. The size of these objects (for the usual degenerate gas configuration) is given by:

$$M^{1/3}R = \frac{92\hbar^2}{Gm_D^{8/3}} \tag{3.24}$$

For a Neptune mass object, the corresponding size will be $R \approx 10^2$ m. For the density of DM of $\sim 0.1$ GeV/cc around the solar neighbourhood, there could be one such object within about half a light year. These DM objects could rotate about their axis, with the limiting speed given by,

$$v = \left(\frac{GM}{R}\right)^{1/2} \tag{3.25}$$

For $M$ and $R$ for the above object, the corresponding frequency and period of rotation are,

$$\omega \approx 1 \, \text{kHz}, \quad P \approx 10^{-3} \, \text{s} \tag{3.26}$$

If these objects are not spherically symmetric and have an ellipticity of, say $\epsilon \approx 0.1$, they would emit gravitational waves, and the energy lost through gravitational waves is again given by,

$$\dot{E} = \frac{32}{5}\frac{G}{c^5}M^2R^4\omega^6\epsilon^2 \tag{3.27}$$

where $M$ and $R$ are the mass and radius of these objects, which follows from Eqs. (3.23) and (3.24). For a typical DM object, as discussed above, of mass $10^{26}$ kg and the size of $10^2$ m, the power lose due to gravitational waves is,

$$\dot{E} \approx 10^{24} \, \text{J/s} \tag{3.28}$$

at a frequency of 1 kHz. This just corresponds to the LIGO frequency. The corresponding flux from a distance of about 100 AU will be, $f \approx 10^{-3}$ J/m$^2$/s. The flux of gravitational waves is related to the strain as:

$$f = \frac{c^3}{32\pi G}\omega^2h^2 \tag{3.29}$$

For a flux of $\approx 10^{-3}$ J/m$^2$/s, the strain is, $h \approx 10^{-22}$. This strain and the frequency ($\sim 1$ kHz) are both within the sensitivity of LIGO.

With about one such object within a solar system volume ($\sim 10^{44}$ m$^3$), and a total galactic DM mass of $\sim 10^{42}$ kg, there could be as much as $\sim 10^{17}$ such DM objects (of course, not all DM particles will coalesce to form these objects). The most massive of these objects ($\sim 10^{-4}\,M_\odot$), i.e. Neptune mass could indeed be candidate for gravitational microlensing observations. We conjecture that there could be one such DM object in a solar system volume. This could be testable. No other such object in our system is expected. As we have noted, the Neptune mass is the upper limit.

There could still be $10^{12}$ such Neptune mass DM objects, either tied to other stellar systems or mostly floating as free planets. The gravitational radiation of around a kilohertz frequency (i.e. within range of LIGO) could arise from such objects, both in individual cases (when the objects are not spherically symmetrical) or in binaries (of such objects) where they have a usual quadrupole moment.

There could be a whole host of such objects starting with asteroid mass (primordial DM planet hypothesis). The halo could consist mainly of such objects with masses $< 10^{-6} \, M_\odot$. Objects of lower mass with similar features or in binaries would also emit gravitational radiation but with reduced frequency, with the frequency scaling with mass as $\nu \propto M$. These could be detectable in future (tuneable) detectors that are being planned. A mass function for the DM objects could be of the form,

$$N(M, dM) = \frac{dM}{M^2}\beta \tag{3.30}$$

where $N(M, dM)$ is the number of such objects between $M$ and $dM$, $\beta$ is a constant. The lowest mass, as estimated above, is $\sim 10^{-18} \, M_\odot$. This power law distribution would put only a small fraction ($\sim 0.1\%$) in the upper mass range of $10^{-4} \, M_\odot$.

We also note that $10^{17}$ objects is an upper limit, assuming all DM is in the form of these objects. Only a fraction of the DM particles may be in the form of these objects, so we have a real upper limit. This will give total gravitational power from all the DM objects of,

$$\dot{E} \approx 10^{41} \, \text{J/s} \tag{3.31}$$

The energy density of gravitational waves in the galaxy will be given by the background flux, $f = \frac{\dot{E}}{4\pi d^2} \approx 10^{-2} \, \text{J/m}^2/\text{s}$, where $d \approx 10^{21}$ m is the distance to the galactic centre.

These DM objects in binaries could also be a source of gravitational waves. If a binary system containing such objects of mass and size given from Eqs. (3.23) and (3.24) are separated by distance, $r = 10R$, then the period is given as:

$$GMP^2 = 4\pi^2 r^3 \tag{3.32}$$

This gives a period of $P \approx 10^{-3}$ s, and frequency $\omega \approx 1$ kHz. The power radiated through gravitational waves would be,

$$\dot{E} = \frac{32}{5}\frac{G}{c^5}M^2 r^4 \omega^6 \approx 10^{26} \, \text{J/s} \tag{3.33}$$

The energy released during the final stages of the merger will be their binding energy, $E = \frac{GM^2}{R} \approx 10^{40}$ J. All the binding energy need not be emitted as gravitational radiation, and this is, again, the upper limit. But in the case of these DM objects, unlike neutron stars, energy is not carried away in the form of neutrinos or electromagnetic waves; hence most of the binding energy would be converted to gravitational radiation.

These binaries merging at around the galactic centre with a millisecond burst of gravitational waves will produce a strain given by,

$$h = \frac{GE}{c^4 r} \sim 10^{-24} \tag{3.34}$$

which is just outside the sensitivity of LIGO. If such a binary merger occurs over a kiloparsec distance, it will give a strain within LIGO sensitivity ($\sim 10^{-22}$), at 1 kHz frequency.

For individual objects and binaries, separated by a distance comparable to their radius, the angular frequencies are $\sim$ kHz (similar to merging neutron stars). The frequency scales with mass $M$, so even a mass ten times smaller than the upper limiting mass would have angular frequency $\sim 100$ Hz. These will be in the range of LIGO. When they merge and collapse, if a black hole is formed, the radius of the black hole is $\sim 10^{-2}$ m, so the ring down time, in this case, would be about a nanosecond. However, if the merging objects are well below the 'Neptune' limiting mass, they would result in the formation of a compact DM object close to limiting mass, so the ring down frequency is now $\sim$ kHz, similar to merging neutron stars (or stellar mass black holes).

Like in the case of black hole mergers, gravitational waves from these dark matter objects will not be accompanied by electromagnetic radiation since dark matter does not couple with radiation. This is unlike the case of the merger of neutron stars. Neutron star merger will produce millisecond bursts of $10^{45}$ J of gravitational waves accompanied by electromagnetic radiation, while the merger of DM objects will produce millisecond bursts of $10^{33}$ J, without corresponding electromagnetic radiation being emitted.

# Chapter 4
# Thermal Gravitational Waves

**Abstract** High-frequency (thermal) gravitational waves from stellar cores, jets, gamma-ray bursts, primordial black holes, etc., are not often discussed. Here we give a detailed account of the physics involved and give estimates of such emissions from the above-mentioned sources. We estimate the backgrounds from these sources. We expect gravitational waves from the early universe going back to the Planck epoch, including from inflation. We estimate the redshifted integrated thermal background gravitational waves from these sources.

Thermal gravitational waves are a theoretical type of gravitational waves that are produced by the thermal motion of matter. These waves are extremely small and difficult to detect, but they are predicted by the theory of general relativity. These waves are predicted to arise due to the random motion of particles in a system, such as the motion of atoms in a gas or the motion of stars in a galaxy. Thermal gravitational waves are different from the gravitational waves that have been detected by LIGO and Virgo, which are generated by the acceleration of massive objects, such as black holes or neutron stars. Thermal gravitational waves would have a very high frequency and would be extremely difficult to detect with current technology. However, the existence of these waves could provide valuable insight into the behaviour of matter at very small scales.

Due to Coulomb collisions in the core of the stars, thermal gravitational waves can be generated. These thermal gravitational waves can arise in white dwarfs and neutron stars due to the fermion collisions in the dense degenerate Fermi gas. Such high frequency thermal gravitational waves are also produced during collisions in a gamma-ray burst or in the jets of a rotating black hole.

## 4.1 Thermal Gravitational Waves from Stellar Cores

Stellar cores are the central regions of stars, where the majority of the stars' mass is concentrated. The hottest and densest part of the star, the core, is typically made up of

A. Kenath and C. Sivaram, *Physics of Gravitational Waves*,
SpringerBriefs in Physics,
https://doi.org/10.1007/978-3-031-30463-7_4

hydrogen and helium. If $n_1, n_2$ are the number densities of these particles undergoing collision with a differential scattering cross-section of $\frac{d\sigma_{12}}{d\Omega}$, with relative velocity $v_{12}$ and reduced mass $\mu_{12}$, then the power per unit volume per unit frequency interval is given by the quadrupole formula as:

$$\dot{E} = \left( \frac{32G}{5c^5} \mu_{12}^2 n_1 n_2 v_{12}^5 \sum \frac{d\sigma_{12}}{d\Omega} \sin^2 \theta \right) V \nu \qquad (4.1)$$

where $V$ is the volume of the stellar core and $\nu = \frac{kT_C}{h} \approx 10^{17}$ Hz is the frequency corresponding to the core temperature of the star of $T_C \approx 10^7$ K.

The velocity of the particles at this core is $v_{12} = \sqrt{\frac{3kT_C}{\mu_{12}}} \approx 8 \times 10^5$ ms$^{-1}$. For a star of density $\rho \approx 2 \times 10^5$ kgm$^{-3}$, the number density $n_1 = n_2 = \frac{\rho}{m_P} \approx 10^{32}$ m$^{-3}$. And the volume of the star is of the order of $V \approx 10^{27}$ m$^3$.

For a main sequence star,

$$\sum \frac{d\sigma_{12}}{d\Omega} \sin^2 \theta = \frac{e^4}{(8\pi\varepsilon_0)^2 \mu_{12}^2 v_{12}^4} \approx 5 \times 10^{-28} \text{ m}^2 \qquad (4.2)$$

Using these values in Eq. (4.1), we get the power of thermal gravitational waves emitted as $\dot{E} \approx 10^9$ W at a frequency of $\nu \approx 10^{17}$ Hz. The flux of thermal gravitational waves from the Sun received at Earth is of the order of half a watt.

## 4.2   Thermal Gravitational Waves from White Dwarfs and Neutron Stars

Compact stars—white dwarfs and neutron stars—are incredibly dense objects that are formed from the collapsed core of a massive star. White dwarfs are the end stages of stars that are below the Chandrasekhar mass limit. They are typically about the size of Earth but have a mass similar to that of the Sun. They are composed mostly of carbon and oxygen. In the case of white dwarfs, the number density is of the order of $n_1 = n_2 \approx 10^{37}$ m$^{-3}$, and the velocity corresponding to the white dwarf temperature of $T_C \approx 10^8$ K, is of the order of $\sim 2 \times 10^6$ m/s. The volume of the white dwarf is of the order of $\approx 4 \times 10^{18}$ m$^3$ and the frequency corresponding to the temperature, $T_C$ is $\nu = \frac{kT_C}{h} \approx 10^{18}$ Hz. And,

$$\sum \frac{d\sigma_{12}}{d\Omega} \sin^2 \theta = \frac{e^4}{(8\pi\varepsilon_0)^2 \mu_{12}^2 v_{12}^4} \approx 10^{-29} \text{ m}^2 \qquad (4.3)$$

The power of thermal gravitational waves emitted by the white dwarf works out to be of the order of $\dot{E} \approx 10^{12}$ W at a frequency of $\nu \approx 10^{18}$ Hz.

When the mass of the stars exceeds the Chandrasekhar limit, they end up as neutron stars. Neutron stars are incredibly dense, with a mass that is around 1.4 times that of the Sun but a diameter of only around 10 km. The number density is of the order of $n_1 = n_2 \approx 10^{44}\,\text{m}^{-3}$, and the velocity corresponding to the neutron star temperature of $T_C \approx 5 \times 10^8\,\text{K}$, is of the order of $\approx 5 \times 10^7\,\text{m/s}$. The volume of the neutron star is of the order of $4 \times 10^{12}\,\text{m}^3$ and the frequency corresponding to the temperature $T_C$ is $\nu = \frac{kT_C}{h} \approx 10^{21}\,\text{Hz}$. And

$$\sum \frac{d\sigma_{12}}{d\Omega} \sin^2\theta = \frac{e^4}{(8\pi\varepsilon_0)^2 \mu_{12}^2 v_{12}^4} \approx 10^{-33}\,\text{m}^2 \tag{4.4}$$

The power of thermal gravitational waves emitted by the neutron star works out to be of the order of $\dot{E} \approx 10^{22}\,\text{W}$ at a frequency of $\nu \approx 10^{21}\,\text{Hz}$.

We assume collisions of neutrons described by the hard sphere fermion model with scattering length of the order of $5 \times 10^{16}\,\text{m}$. We restrict to S-wave scattering since the de Broglie wavelength is large compared to this length. The integrated power density is given by:

$$P_g = \frac{8G}{5c^5}(3\pi^2 n)^{2/3} l^2 \left(\frac{M}{m_n}\right)\left(\frac{kT_C}{\hbar^{1/2}}\right)^4 \tag{4.5}$$

where $M$ is the mass of the star, $m_n$ is the neutron mass, $n$ is the central number density and $T_C$ is the core temperature assumed to be much smaller than the Debye temperature, $T_D\, (= 10^{14}\,\text{K})$. Even for a newly formed hottest neutron star $T_C = 5 \times 10^{10}\,\text{K} < T_D$.

## 4.3 Thermal Gravitational Waves from Gamma-Ray Bursts

Gamma-ray bursts (GRBs) are the most luminous physical phenomena in the universe known to the field of astronomy. They consist of flashes of gamma-rays that last from seconds to hours, the longer ones being followed by several days of X-ray afterglow. Similar to the stars, in GRBs also, Coulomb collisions can result in the emission of high frequency gravitational waves.

The power of the thermal gravitational waves is given by the same expression as that for the stars, but the bulk properties will be altered by factors of gamma ($\Gamma$) due to the relativistic velocities encountered in GRBs. The number density will be increased by a factor of gamma, and the volume associated with the GRB will be given by the deceleration volume. In gamma-ray bursts, due to the general relativistic effects, the shock wave propagated from the burst will be decelerated. The blast wave will form a spherical shell around the blast. The radius of this shell is called the deceleration radius.

To determine this consider the energy of GRB given by,

$$E_v = \frac{4}{3}\pi R_D^3 \Gamma^2 n m_p c^2 \tag{4.6}$$

Hence the volume is given by:

$$V = \frac{E_v}{\Gamma^2 n m_p c^2} \tag{4.7}$$

For $n = 10^8$ m$^{-3}$, the volume is of the order of $V = \frac{5 \times 10^{46}}{\Gamma^2}$ m$^3$.
This gives the power of the thermal gravitational waves emitted from a GRB as,

$$\dot{E} = \left( \frac{32G}{5c^5} (\Gamma \mu_{12})^2 \left( \frac{N}{V} \right)^2 v_{12}^5 \sum \frac{d\sigma_{12}}{d\Omega} \sin^2 \theta \right) V v \tag{4.8}$$

Putting in the value for the volume, $n$, and $\frac{d\sigma_{12}}{d\Omega}$, we get, $\dot{E} = \Gamma^4 (3.5 \times 10^8)$ W.
A gamma-ray burst corresponding to $\Gamma = 100$, and the power works out to be of the
order of $3.5 \times 10^{16}$ W at the frequency $v = \Gamma\left( \frac{kT}{h} \right) \approx 10^{22}$ Hz.

## 4.4  Thermal Gravitational Waves from Short Duration GRBs

Short gamma-ray bursts have a shorter duration ($< 0.2$–2 s) and a harder spectrum as
compared to the duration of 2–200 s for long GRBs. The first of these short GRBs
(GRB050509b) was identified with the halo of an elliptical galaxy at a distance of
1.12 Gpc. Short GRBs are due to the merger of two neutron stars, whereas long
GRBs are due to the collapse of very massive stars. The spectrum observed is harder
because the objects merging to produce the GRB are more compact. The time taken
for the merging of two NS is given by,

$$\tau_{merger} = \frac{GM^2}{R} / \dot{E} \tag{4.9}$$

Here, $\dot{E}$ is the loss of energy due to the emission of thermal gravitational waves.
For two neutron stars of mass 1.5 $M_\odot$ each, the merger time typically works out
to be of the order of $10^9$ years. Due to the longer merger time of the neutron stars, the
short GRBs are found in older population elliptical galaxies. In the case of the merger
of two neutron stars, the number density, as well as the temperature, is substantially
high compared to the long duration GRB, with the temperature of the order of $10^{13}$ K
and $n = 10^{46}$ m$^{-3}$. Considering all the gamma factors associated with the GRB, as in
the previous case, the gravitational power for the short GRB is, $\dot{E} = \Gamma^4 (4 \times 10^{36})$ W,
at a frequency $v = \Gamma\left( \frac{kT}{h} \right) \approx \Gamma(10^{23})$ Hz.

During the short duration burst, as the two neutron stars undergo collision, their tidal breakup releases its binding energy. For a neutron star with a mass of 1.5 $M_\odot$ and radius of about 10 km, the binding energy released is,

$$BE = 2\left(\frac{3}{5}\frac{GM_{NS}^2}{R_{NS}}\right) \approx 6 \times 10^{46}\,\text{J} \tag{4.10}$$

For a gamma factor of about 100, the power radiated due to the thermal gravitational wave emission is of the order of $4 \times 10^{44}$ W. This implies that about one per cent of the energy released in the short duration gamma-ray burst could be in the form of thermal gravitational waves. For a typical distance of 100 Mpc for the GRB, the flux is given by,

$$f = \frac{E}{4\pi d^2} \approx 4 \times 10^{-6}\,\text{W/m}^2 \tag{4.11}$$

Since the event occurs over a time scale of one second, the flux is equivalent to the fluence.

## 4.5 Thermal Gravitational Waves from Jets

The effect of the spin of the black hole (Kerr back holes) produces a very interesting phenomenon of powerful bi-directional jets. The process involves a supermassive black hole that is continuously fed with magnetized gas through an orbiting accretion disk. The combination of the strong gravity of the black hole, the rotation in the infalling matter, and the magnetic field are responsible for the jet creation. The jet is powered by both the energy of accretion in the disk and from the rotational energy of the black hole. It is the rotating black hole, however, that provides most of the energy.

The length of the jet in terms of the mass of the central black hole and the number density of the ambient gas is given by, $l = \left(\frac{3GM^2}{\pi\rho c^2(\tan 5)^2}\right)^{1/4}$. The corresponding volume of the jet in terms of the mass of the central black hole and the number density of the ambient gas is,

$$V = \frac{1}{\Gamma}\left(\frac{M}{\sqrt{n}}\right)^{3/2} \tag{4.12}$$

And the power associated with the thermal gravitational waves from the jet is given by,

$$\dot{E} = \left( \frac{32G}{5c^5} (\Gamma \mu_{12})^2 \left( \frac{N}{\frac{1}{\Gamma} \left( \frac{M}{\sqrt{n}} \right)^{3/2}} \right)^2 v_{12}^5 \left( \frac{e^4}{(8\pi \varepsilon_0)^2 \mu_{12}^5 v_{12}^4} \right) \right) \left( \frac{1}{\Gamma} \left( \frac{M}{\sqrt{n}} \right)^{3/2} \right) v$$

(4.13)

For a given black hole, the length of the jet depends on the density of the particles emitted along the jet.

For a 30 $M_\odot$, with a number density of $n = 10^3$ m$^{-3}$, the length of the jet is of the order of $l \approx 1$ kpc. Using these in Eq. (4.13), the thermal gravitational waves from the jets will be, $\dot{E} = \Gamma^3 (6.5 \times 10^{20})$ J/s. Here we notice a $\Gamma^3$ dependence in the case of the thermal gravitational waves from a jet, where as we had a $\Gamma^4$ dependence in the case of gamma-ray bursts.

## 4.6  Thermal Gravitational Waves from PBHs

Primordial black holes (PBHs) are a hypothetical type of black holes that are formed not by the gravitational collapse of a star but by the extreme densities of matter present during the early universe, shortly after the Big Bang. These black holes are thought to be much smaller than the more common black holes that are formed from the collapse of stars. The existence of primordial black holes was first proposed by Stephen Hawking in the 1970s. He suggested that smaller black holes could have been created in the dense, high-energy environment of the early universe. These black holes would have been so small that they would have evaded detection.

Over the years, there have been many theories and predictions about the properties of primordial black holes. One of the key challenges in studying primordial black holes is the fact that they are so small and difficult to detect. However, recent advances in gravitational wave technology have made it possible to search for these elusive objects. Although classical black holes are 'black', Hawking suggested that incorporating quantum effects can lead to black holes radiating. This is the so-called Hawking radiation, a phenomenon where black holes radiate energy, resulting in the loss of mass and, eventually, the black hole's evaporation. Hawking radiation has been observed in the lab using analogue systems, but it has not yet been directly observed in a black hole.

As these PBHs evaporate through Hawking radiation, part of the energy could be released in the form of thermal gravitational waves. In the case of spontaneous graviton emission, the quadrupole gravitational power is given by,

$$P_{GW} = \frac{G}{c^5} m^2 \omega^6 R^4$$

(4.14)

The evaporation time for a PBH is given by, $t_{PBH} = \frac{G^2 M^3}{\hbar c^4}$. A typical PBH that evaporates over the Hubble time has a mass $m \approx 10^{11}$ kg and the corresponding

Schwarzschild radius, $R_S \approx 10^{-15}$ m. The power due to the gravitational waves can be written as,

$$P_{GW} = \frac{G}{c^5} \left( \frac{mv^2}{t} \right)^2 \tag{4.15}$$

where $v$ is the typical velocity and $t$ is the time scale of the underlying explosive processes giving rise to the gravitational wave emission. The term within the bracket in Eq. (4.15) corresponds to the power of the explosion; therefore, we have,

$$P_{GW} = \frac{G}{c^5} P_{exp}^2 \tag{4.16}$$

where $P_{exp}$ is the explosion power. The total energy released in the form of gravitational waves is, $E_{GW} = \left( \frac{G}{c^5} P_{exp}^2 \right) t$. The number of quanta of gravitational waves is then given by,

$$N_{GW} = \frac{E_{GW}}{\hbar/t} = \frac{G}{c^5} \frac{P_{exp}^2 t^2}{\hbar} = \frac{G}{c^5} \frac{E_{exp}^2}{\hbar} \tag{4.17}$$

where, $\hbar/t$ is the energy of each gravitational wave quanta. For the typical PBH of mass of $10^{11}$ kg, the energy of the explosion is $E_{exp} = mc^2 \approx 10^{28}$ J. Hence the quanta of gravitational waves (from Eq. (4.17)) are, $N_{GW} \approx 10^{38}$. The frequency and the corresponding energy of the gravitational wave quanta are given by, $v = 1/t$ and $\varepsilon = \hbar/t$, respectively. Here, $t = R_S/c \approx 10^{-23}$ s. Therefore each quanta of gravitational waves have an energy of $\approx 10^{-11}$ J ($\sim 100$ MeV) with a frequency of $\approx 10^{23}$ Hz.

The total energy associated with the gravitational waves is then, $E_{GW} = \varepsilon N_{GW} \approx 10^{27}$ J. As we see from above, about 10% of the energy of the PBH explosion is converted to high frequency gravitational waves, with a typical frequency of $\approx 10^{23}$ Hz corresponding to $\sim 100$ MeV.

According to current estimates, the number density of PBHs in the universe could be $\sim 1/(\text{kpc})^3$. Then the total energy associated with the gravitational waves from these PBHs over the entire volume of the universe of $\left( 10^{26} \text{ m} \right)^3$ is, $E_{total} \approx 10^{47}$ J. This corresponds to an energy density of $\approx 10^{-31}$ J/m$^3$, and the corresponding flux of,

$$f_{total} \approx 10^{-23} \text{ J/m}^2/\text{s} \tag{4.18}$$

in high energy thermal gravitational waves at typical energies of $\sim 100$ MeV. If the number density of PBH is $\sim 1/(\text{pc})^3$, then the flux would be, $f_{total} \approx 10^{-13}$ J/m$^2$/s.

The above discussion pertains to only those PBHs that evaporate over the Hubble time of $\approx 10^{10}$ years. The integrated flux from all the PBHs over the entire Hubble time will give the background thermal gravitational flux. The number of the PBHs as a function of mass can be written as,

$$n_{BH}(m) = n_{BH}(m_0)\left(\frac{m_0}{m}\right)^{-n} \tag{4.19}$$

where $n_{BH}(m_0)$ is the present number density of PBHs, $m_0$ is the mass of the PBH that evaporate over the Hubble time and $n = 3$. The integrated energy of the gravitational wave emission over the entire volume of the universe is given by,

$$E_{integrated} = \int\limits_{m_1}^{m_2} \frac{m^2 c^4}{\hbar} \frac{c}{H_0} n_{BH}(m_0)\left(\frac{m_0}{m}\right)^{-n} dm \times V \tag{4.20}$$

The upper limit of the integral, $m_2 = 10^{11}$ kg, is the mass of the PBH that will evaporate over the Hubble time scale and the lower limit, $m_1 \approx 10^{10}$ kg, is the mass limit below which the flux is too small. The volume $V = 2\pi^2 R_H^3$, where $R_H = 10^{26}$ m is the Hubble radius. From these values, we get the total integrated energy as, $E_{integrated} \approx 10^{25}$ J, and the total integrated energy flux as, $f_{integrated} \approx \frac{E_{integrated}}{4\pi R_H^2} \approx 10^{-27}$ J/m$^2$/s.

The energy associated with each of the thermal gravitational wave quanta is of the order of $\approx 10^{-11}$ J. Therefore the total flux of thermal gravitational wave quanta is, $N_{integrated} \approx 10^{-16}$/m$^2$/s. If the number density of the PBHs is $\sim 1/(\text{pc})^3$, then the flux would be of the order of $10^{-14}$ J/m$^2$/s.

## 4.7  Background Thermal Gravitational Radiation

Quantum considerations would imply that the earliest epoch of the universe would have been the Planck epoch starting at the time given by, $\left(\frac{\hbar G}{c^5}\right)^{1/2} \approx 10^{-44}$ s. This is the minimal time given from quantum considerations. At this Planck epoch, all the interactions were of equal strength so that thermal equilibrium was maintained between gravitons and other particles. As the universe expanded, the gravitational interaction weakened and graviton decoupled from other particles. If $N$ is the number of particles that were coupled with the graviton, then the temperature of the background gravitational radiation is given by,

$$T_{GW} = \left(\frac{43}{22N}\right)^{1/3} T_{rad} \tag{4.21}$$

where, $T_{rad}$ is the background radiation temperature. And for the present radiation temperature of about 2.7 K and $N \approx 30$, we have the temperature of the background gravitational radiation of 1 K.

The implication of this result is in the confirmation of the inflationary model of the universe, according to which the universe went through an exponential expansion of $R \approx \exp\left(\sqrt{\Lambda t}\right)$. At the time of inflation, the expansion of the universe occurred at an exponential rate, where the expansion was by a factor of $10^{28}$. The expansion time of the universe when inflation occurred is about $10^{-36}$ s. The radiation temperature at this epoch is given by:

$$T_{rad} = \left( \frac{3c^2}{32\pi Ga} \right) \frac{1}{\sqrt{t}} \approx \frac{2 \times 10^{10}}{\sqrt{t(s)}} \approx 2 \times 10^{28} \text{ K} \qquad (4.22)$$

The temperature of the background thermal gravitational radiation (from Eq. (4.21)) corresponding to this radiation temperature is then, $T_{GW} \approx 8 \times 10^{27}$ K. And the corresponding wavelength is:

$$\lambda = \frac{\hbar c}{k T_{GW}} \approx 10^{-30} \text{ m} \qquad (4.23)$$

At the end of the inflation phase, the wavelength of the gravitational radiation background is $10^{-2}$ m. However, in order not to interfere with nucleosynthesis in the hot dense phase, its energy density would have to be less than one per cent of the radiation energy density. This would give a strain of the order of $(\hbar\omega c)^2/G$. At the present epoch, the wave would be stretched by a further factor of $10^{19}$. This would give a wavelength of $10^{17}$ m at present. Detection of such waves through fluctuations in the cosmic microwave background radiation could verify the existence of such a phase in the early universe.

If inflation had not taken place, we would be left with a thermal gravitational wave background with a temperature of 1 K. If at all this can be detected, it would provide evidence against inflation. Also, the detection of these thermal background gravitational waves provides a basis to verify the validity of the big bang model itself. There are claims that the most convincing evidence of the big bang, the microwave background and abundance of helium, can be accounted for without invoking the big bang. But these thermal gravitational waves cannot be generated without the universe passing through the super-hot, super dense Planck epoch.

Interestingly enough, Eq. (4.20) can be used to put a limit on any primordial intergalactic magnetic field. For instance, if the primordial magnetic field $\sim 0.1$ G at recombination, i.e. at a redshift $z \approx 10^3$, then the flux $f \approx 10^{-3}$. Larger values of $f$ would cause anisotropies in the photon background larger than measured and would result in a noticeable weakening of a single polarization of the microwaves. This already limits present-day magnetic fields to less than $10^{-6}$ G in intergalactic space, or less than $10^{-4}$ G in interstellar space. The number density of the graviton background is of the order of $10^9$ m$^{-3}$ so that their flow through a square kilometre area is of the order of $10^{24}$ s$^{-1}$.

The thermal gravitational waves continue to be generated in the early universe as it expands, as the temperature and particle densities continue to be high. We can estimate the integrated power emitted in thermal gravitational waves as the universe cools from (say) $10^{13}-10^5$ K. This is similar to the discussions on the stellar core, but now all the quantities are time dependent. The time range corresponding to the above temperatures is $10^{-6}-10^{10}$ s.

The power per unit volume per unit frequency interval is given by the quadrupole formula as given in Eq. (4.1), and the differential cross section is given by Eq. (4.2). The number density is dependent on time as $n \propto t^{-2}$, which implies,

$$n = n_i \left(\frac{t_i}{t}\right)^2 \tag{4.24}$$

Here the quantities with the suffix '$i$' indicate the initial values of these quantities at the time $10^{-6}$ s. The frequency and the velocity are given by, $\nu = kT/\hbar$ and $\nu_{12} = \sqrt{3kT/\mu_{12}}$ respectively. The temperature dependence on time is given by, $T = \frac{10^{10}}{\sqrt{t(s)}}$. Hence the time dependence of frequency and velocity is, $\nu = \nu_i \left(\frac{t_i}{t}\right)^{1/2}$ and $\upsilon = \upsilon_i \left(\frac{t_i}{t}\right)^{1/4}$. The radius of the universe is related to the temperature as $RT$ is a constant. Hence we have the volume dependence given by,

$$V = V_i \left(\frac{T_i}{T}\right)^3 = V_i \left(\frac{t_i}{t}\right)^{3/2} \tag{4.25}$$

Using these results in the expression for power radiated by the thermal background gravitational radiation we have,

$$\dot{E} = \left(\frac{32G}{5c^5}\right) n_i^2 \nu_i \left(\frac{e^4}{(8\pi\varepsilon_0)^2}\right) V_i \upsilon_i \left(\frac{t_i}{t}\right)^{3/4} \tag{4.26}$$

The initial values corresponding to time $t = 10^{-6}$ s is given by,

$$n = n_i \left(\frac{t_i}{t}\right)^2 = 1 \text{ proton/m}^3 \left(\frac{T_H}{10^{-6}}\right)^2 \tag{4.27}$$

$$\nu_i = kT_i/\hbar \approx 2 \times 10^{23} \text{ Hz and } \upsilon_i = \sqrt{3kT_i/m_p} \approx 3 \times 10^8 \text{ m/s} \tag{4.28}$$

$$R_i = R\left(\frac{T}{T_i}\right) = cT_H\left(\frac{2.7}{10^{13}}\right) \text{ and } V_i = 2\pi^2 R_i^3 \approx 10^{43} \text{ m}^3 \tag{4.29}$$

Using Eqs. (4.27)–(4.29) in Eq. (4.26), we get,

$$\dot{E} = \frac{2 \times 10^{56}}{t^{3/4}} \tag{4.30}$$

The integrated energy is then given by,

$$\int \dot{E}dt = \int_{10^{-6}}^{10^{10}} \left(\frac{2 \times 10^{56}}{t^{3/4}}\right) dt \approx 10^{60} \text{ J} \tag{4.31}$$

This is the integrated energy radiated by the background thermal gravitational radiation in the early universe as the temperature cooled from $10^{13}-10^5$ K. As the universe expanded, this energy is red shifted by a factor given by,

$$\frac{R\text{ (present)}}{R\left(\text{at }T = 10^5\text{ K}\right)} \approx 10^5 \tag{4.32}$$

Therefore the present energy associated with the integrated background thermal gravitational radiation is of the order of $10^{55}$ J. If $E/V_{\text{present}}$ is the energy density, where, $V_{\text{present}} = 4 \times 10^{79}$ m$^3$ is the present volume of the universe, then the flux associated with this integrated background thermal gravitational radiation is given by, $f = 2 \times 10^{-17}$ J/m$^2$/s.

The energy density associated with the cosmic microwave background radiation is of the order of $10^{-14}$ J/m$^3$, and the corresponding flux is, $f_{cmbr} \approx 10^{-6}$ J/m$^2$/s. We see that the flux associated with integrated background thermal gravitational radiation is about ten orders less than that of the CMB radiation. The contribution of the microwave background radiation to the normalised critical density of the universe is $\Omega_{cmbr} \approx 4 \times 10^{-5}$. Since the flux associated with the integrated background thermal gravitational radiation is about ten orders less, the contribution to the normalised critical density of the universe due to this will be $\Omega_{ibtgr} \approx 10^{-15}$.

Added to this would be the contribution to the background thermal gravitational radiation by the astrophysical objects, as discussed in sections 4.1 through 4.6. The power of thermal gravitational waves emitted by a star is of the order of $10^9$ W. Considering all the $10^{11}$ stars in the $10^{11}$ galaxies, the power associated with the thermal gravitational waves from all these stars is of the order of, $10^{31}$ W. The total energy emitted due to the thermal gravitational waves by these stars over their average life span of $\approx 3 \times 10^{17}$ s is, $\approx 3 \times 10^{48}$ J.

The power of thermal gravitational waves emitted by neutron stars is of the order of $10^{22}$ W. If one out of every 100 stars is a neutron star, then the power associated with the thermal gravitational waves from all the neutron stars is, $\approx 10^{42}$ W, and the total energy emitted due to the thermal gravitational waves by these neutron stars over their average life span of $\approx 10^{18}$ s is, $\approx 10^{50}$ J.

As for the PBHs, we need something like $10^{30}$ of them to match the thermal gravitational wave background. This would need a density of such objects much more than what is implied by the gamma-ray background. The overall energy associated with the emission of thermal gravitational waves from stellar sources is of the order of $10^{50}$ J, which is still about five orders less than that associated with integrated background thermal gravitational radiation.

# Chapter 5
# Detection of Gravitational Waves

**Abstract** We mention the various methods invoked to detect gravitational waves, starting from Weber bar detectors. Details of the current LIGO detectors and their results are discussed. We give an overview of future detectors like LISA and MIGA. These detection methods are not viable for high frequency gravitational waves. The various suggestions which have been made to detect thermal and other high frequency gravitational waves are elaborated.

Though the Hulse–Taylor observations were very important, they give only indirect evidence for gravitational waves. The more conclusive observation came with the direct measurement of the effect of a passing gravitational wave, which also provides more information about the system that generated it. Such direct detections are complicated by the extraordinarily small effect the waves would produce on detectors. It is extremely difficult to generate gravitational waves in the lab, and there is no Hertz experiment equivalent so far in the case of gravitational radiation. For instance, a rod one metre long, spinning close to breaking speed, would only produce a power of $\sim 10^{-32}$ W, which is too weak to be detected.

Weber bar is a simple device to detect gravitational waves. It is a large, solid bar made of a metal alloy, such as tungsten and isolated from outside vibrations. Strains in space due to an incident gravitational wave excite the bar's resonant frequency and could thus be amplified to detectable levels. Nearby supernovae might be strong enough to be seen without resonant amplification. Modern forms of the Weber bar are still operated, cryogenically cooled, with superconducting quantum interference devices to detect vibration (for example, ALLEGRO). Weber bars are not sensitive enough to detect anything but extremely powerful gravitational waves.

## 5.1 Laser Interferometer Gravitational-Wave Observatory

Laser Interferometer Gravitational-Wave Observatory (LIGO) is a large-scale experiment tasked with the direct detection of gravitational waves. It was co-founded in

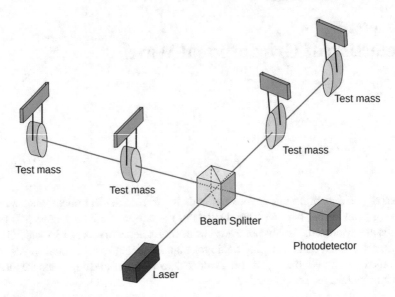

**Fig. 5.1** LIGO gravitational wave detector

1992 by Kip Thorne and Ronald Drever of Caltech, and Rainer Weiss of MIT. LIGO is a joint project between MIT, Caltech, and many other colleges and universities and is sponsored by the National Science Foundation, USA, with the observations starting in 2002 (Fig. 5.1).

The original detectors were disassembled and were replaced by improved versions known as "Advanced LIGO". The observatory consists of two detectors, one in Hanford, Washington and the other in Livingston, Louisiana. Each detector is a four kilometre long L-shaped vacuum tube with extremely sensitive instruments at the intersection. When a gravitational wave passes through Earth, it stretches and compresses the length of the detectors, causing a tiny change in the distance between the instruments.

For LIGO, with an arm length of four kilometres, a mass of end mirror $M \approx 1\,\text{T}$, and $\tau = 1\,\text{ms}$, we have, $h_{\min} \approx 10^{-23}$. This is the sensitivity of the device. Longer the effective arm length (i.e. more multiple reflections), the lower (more sensitive) the value of strain $h$.

## 5.2   Laser Interferometer Space Antenna

The Laser Interferometer Space Antenna (LISA) is a proposed space mission concept designed to detect and accurately measure gravitational waves. The LISA Project's present incarnation is the Evolved Laser Interferometer Space Antenna (eLISA) after NASA pulled out from the partnership with the European Space Agency. The tentative launch date is in 2037. LISA has a constellation of three spacecraft arranged

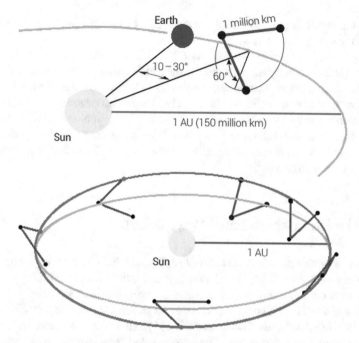

**Fig. 5.2** The three eLISA spacecrafts that follow the Earth in a triangle formation

in an equilateral triangle with million-kilometre arms flying along an Earth-like heliocentric orbit (Fig. 5.2).

The lasers will be used to measure the distance between the spacecraft to an extraordinary level of precision, allowing LISA to detect the tiny changes in distance that are caused by gravitational waves. Unlike gravitational wave observatories on Earth, LISA's arms cannot be held in place at a fixed length. Instead, the distances between its satellites vary greatly over the course of a year's orbit. The detector must continuously track these changing distances, measuring the millions of wavelengths that the distances change each second.

## 5.3 Matter-Wave Laser Interferometric Gravitation Antenna

The Matter-wave laser Interferometric Gravitation Antenna (MIGA) is an underground instrument installed in the Low Noise underground Laboratory (LSBB) in the South-East of France, located away from major anthropogenic disturbances and benefitting from very low background noise. MIGA combines atom and laser interferometry techniques, manipulating an array of atomic ensembles distributed along the antenna to simultaneously read out seismic effects, inertial effects and, eventually, the passage of a gravity wave.

This gravitational waves detector has a 150 m long horizontal base and employs cold-atom interferometry to make precise measurements of variations in gravity and strain. It includes three interferometers that are spaced apart and powered by two lasers that travel in opposite directions. The device can measure the output state of the interferometer by using a detection system that combines velocity-selective Raman spectroscopy with fluorescence detection that is sensitive to a specific type of atomic state. It can achieve a signal-to-noise ratio of 1000 with atoms in the state needed for the interferometry.

## 5.4   Kamioka Gravitational Wave Detector

Kamioka Gravitational Wave Detector (KAGRA) is a large scale cryogenic gravitational wave telescope. It was developed by the Institute for Cosmic Ray Research (ICRR) with contributions from the National Astronomical Observatory of Japan (NAOJ) and the High Energy Accelerator Research Organization (KEK) and is located in Kamioka, Japan. This detector is a Michelson interferometer with the two arms extending three kilometres. They are isolated from external disturbances by suspending its mirrors and instrumentation, and its laser beam operates in a vacuum.

Cryogenic Laser Interferometer Observatory (CLIO) is a prototype detector for gravitational waves situated 1000 m underground in the Kamioka Observatory. It is testing cryogenic mirror technologies for KAGRA. It is an optical interferometer with two perpendicular arms, each having a length of 100 m. The mirrors are cooled to 20 K, which reduces various thermal noise sources that usually affect other gravitational wave observations.

## 5.5   Detection of Thermal Gravitational Waves

When a weak gravitational wave passes through a gas along a line perpendicular to the plane of the particles, then the particles will oscillate. The area enclosed by the particles does not change, and there is no motion along the direction of propagation. Passing of gravitational waves through a system of mass sets it into harmonic oscillations, hence causing a strain $h$ of the order of $10^{-22}$. But what about the detection of thermal gravitational waves? So far, there have been few attempts to conceive the detection of such waves. As estimated earlier, the flux of thermal gravitational waves from the Sun (around a frequency of $\approx 10^{16}$ Hz) at Earth is about 0.5 W. How can we detect this kind of high frequency gravitational wave radiation?

For example, in magnetised plasma, gravitational waves can be coupled to electromagnetic waves and can get damped. The damping time is given by,

$$\tau_{damp} = \frac{\omega_B}{GnT^{1/2}m_n^{3/2}} \tag{5.1}$$

For a magnetic field of $10^{15}$ G and temperature of $10^{12}$ K, the damping time is of the order of $10^2$ s. If two particles in the gas are separated by a distance of $d$, and the passing of a gravitational wave causes a strain $h$, then the change in the distance between the particles is $\Delta d = hd$. The distance of separation is given by, $d = \frac{v}{\omega}$ where, $\omega$ is the frequency and $v = \sqrt{kT/m}$ is the velocity. The change in velocity due to the passing of the gravitational wave is $\Delta v = hd\omega$.

The energy change per collision is given by:

$$dE = \frac{1}{2}m\left(\frac{\Delta v}{v}\right)^2 \tag{5.2}$$

If the number density of the gas is $n$ and $\lambda$ is the mean free path, the total power radiated by the volume is given as,

$$\dot{E} = \frac{1}{2}m\left(\frac{\Delta v}{v}\right)^2 \lambda n = \frac{m^2 \lambda n}{2}\frac{(hd\omega)^2}{kT} \tag{5.3}$$

The energy density of the wave is of the order of, $\frac{(h\omega c)^2}{G}$. On integrating the expression for power, we get the time scale for the damping of the wave as:

$$\tau = \lambda\left(\frac{c}{v}\right)^3\left(\frac{\omega^2}{Gmn}\right) = \lambda\left(\frac{c}{v}\right)^3\left(\frac{\omega^2}{\omega_G^2}\right) \tag{5.4}$$

The quantity $Gmn$ has the same dimensions as $\omega^2$, hence $\sqrt{Gmn}$ can be interpreted as the gravitational plasma frequency $\omega_G$ associated with the gas undergoing oscillations due to the passing of gravitational waves. In the case of neutron stars with a temperature of the order of $T \approx 10^{11}$ K, we have, $\omega \approx \omega_G$. The damping time will be of the order of $10^{-18}$ s and hence waves may be trapped within the star. The trajectories of charged particles may be affected by the passage of gravitational waves, which involves the generation of electric current in the magnetised plasma.

The high frequency gravitational waves can also be detected through the atomic transitions induced by them at a very slow rate. The quadrupole transition of hydrogen from $3d \rightarrow 1s$ state with the emission of a graviton occurs at a frequency of $\approx 10^{15}$ Hz. The quantum mechanical transition rate is given by, $\zeta = \frac{P}{\hbar\omega}$, where $P$ is the power emitted by a dipole. For the case of spontaneous graviton emission, the quadrupole gravitational power is given by:

$$P = \frac{2G\omega^6}{5c^5}I^2 \tag{5.5}$$

The normalised wavefunctions for $1s$ and $3d$ states are given by:

$$\psi_{1s} = \frac{1}{\sqrt{\pi}a^{3/2}}e^{-r/a}; \ \psi_{3d} = \frac{1}{162\sqrt{\pi}a^{3/2}}\left(\frac{r}{a}\right)^2 e^{-r/3a}\sin^2\theta \qquad (5.6)$$

where, $a = \frac{\hbar}{m_e e^2}$ is the Bohr radius. Using these expressions in transition rate, we get $\zeta \approx 6 \times 10^{-40}\,\mathrm{s}^{-1}$, and the corresponding lifetime of the transition is, $\tau \approx 10^{36}\,\mathrm{s}$. The frequency of this transition is of the order of $10^{16}$ Hz, which is within the range of thermal gravitons emitted from the Sun. About $10^3\,\mathrm{m}^{-2}\mathrm{s}^{-1}$ gravitons fall on the Earth from the Sun at this frequency. Thus there is a finite probability of detecting induced emission with a sufficiently large detector. This radiation will be very penetrating.

By coincidence, the lifetime for this transition to take place is the same as the proton decay time of $\sim 10^{31}$ years. Proton decay is a major prediction of the Grand Unified Theories (GUT) and, despite the long lifetime, is being tested by several experiments. So it may not altogether be impossible to also observe the effects of high frequency thermal gravitational radiation, which can also induce transitions with a lifetime comparable to that of proton decay. The absorption rate of these high frequency gravitons is estimated to be $\approx 10^{-27}$, for terrestrial detectors, so one requires a detector of several hundred square kilometres to detect a few transitions in some decades.

Another way of detecting these thermal radiations is to convert them into electromagnetic waves of the same frequency. When an electromagnetic wave of amplitude $H_Y$ propagates through a constant magnetic field $H_0$, it produces a quadrupole stress term given by:

$$T_{YY} = H_Y H_0 \cos(kx - \omega t) \qquad (5.7)$$

This stress term gives rise to gravitational waves given by the linear Einstein equation:

$$\Box h_{YY} = kT_{YY}; \ k = \frac{16\pi G}{c^4} \qquad (5.8)$$

Alternatively, a weak gravitational wave $h_{YY}$ propagating through a magnetic field $H_0$ gives rise to a magnetic field perturbation given by:

$$\Box H_Y = \omega^2 h_{YY} H_0 \qquad (5.9)$$

The fraction of the gravitational wave energy converted into electromagnetic waves of frequency $\omega$ is given by, $f = kH_0^2 d^2$, where $d$ is the special extent of the uniform magnetic field.

# Chapter 6
# Some More Aspects of Gravitational Waves

**Abstract** In this chapter, other aspects related to gravitational waves are discussed. These include polarisation as well as possible constraints on dipole radiation predicted by alternate theories. Emissions from pulsars, including millisecond pulsars, are also discussed. Also, applications to future gravitational wave astronomy, such as the independent estimate of Hubble's constant and constraint on cosmic strings, are elaborated.

## 6.1 Polarisation of Gravitational Waves

Another central prediction of general relativity is the existence of only two gravitational-wave polarizations: the tensor plus and cross modes, with spatial strain tensors given by, $\hat{e}_+ = \begin{pmatrix} 1 & 0 & 0 \\ 0 & -1 & 0 \\ 0 & 0 & 0 \end{pmatrix}$; $\hat{e}_\times = \begin{pmatrix} 0 & 1 & 0 \\ 1 & 0 & 0 \\ 0 & 0 & 0 \end{pmatrix}$ (assuming waves propagating in the $+\hat{z}$ direction). Generic metric theories of gravity, however, can allow for up to four additional polarizations: the x and y vector modes and the breathing and longitudinal scalar modes.

The observation of vector or scalar modes would be in direct conflict with general relativity; hence the direct measurement of gravitational-wave polarisations is an avenue to test theories of gravity. The two polarisations correspond to different ways in which the gravitational wave distorts space-time. In the plus polarisation, the distortion is primarily in one direction, while in the cross polarisation, the distortion is at right angles to the first. For the two possible polarisations, $h_+$, $h_\times$, the total strain is $h = \sqrt{h_+^2 + h_\times^2}$.

The gravitational-wave signal GW170814, observed by both the Advanced LIGO and Virgo detectors, favoured the model assuming pure tensor polarisation over models with pure vector or scalar polarisations. The ability of the Advanced LIGO-Virgo network to study the polarisation of gravitational-wave transients is limited by

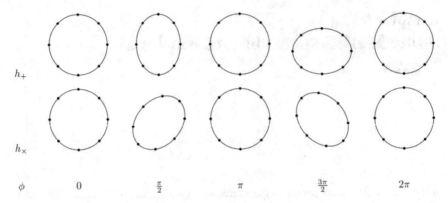

**Fig. 6.1** The effect of the passage of the linearly polarised gravitational wave through the ring of particles in the direction orthogonal to the plane

several factors. At least five detectors are needed to fully characterize the five polarisation degrees of freedom accessible to quadrupole detectors. Quadrupole detectors have degenerate responses to breathing and longitudinal modes and can therefore measure only a single linear combination of scalar breathing and longitudinal polarizations (Fig. 6.1).

## 6.2  Deducing Hubble's Constant from Gravitational Wave Observations

Recently much controversy has been raised about the cosmological conundrum involving the discrepancy in the value of the Hubble constant as implied by Planck satellite observations of the CMBR in the early universe and that deduced from other distance indicators (standard candles like SN, the tip of RG branch, etc.) in the present epoch. The Planck estimate is about 67 km/s/Mpc, while that deduced from distance indicators at the present epoch is around 73–74 km/s/Mpc. The Hubble constant measures the rate of expansion of the universe so that the above values suggest that the universe is expanding faster at present than at the time of the CMBR epoch.

It is implied by Reiss and others that this is a real discrepancy and not due to observational uncertainties. The difference amounts to about ten per cent in the value of the expansion rate, well above the errors and uncertainties. Several suggestions have been made as to the cause of such a discrepancy. Perhaps a new kind of matter or unknown particles has now become more dominant, adding to the 'push' in the expansion rate. For instance, a recent suggestion involves a calculation of the amount of change in the quantum fields needed to account for the dark energy (DE) change. This quantum field causing a change in the DE implies the existence of a new particle with a mass roughly that of the axion (about $10^{-3}$ eV), already predicted earlier. As the background density decreases, this particle is now more dominant.

Indeed that the cosmological constant (presumed to be DE) could be connected to the axion mass was pointed out much earlier. Arguments based on the renormalisation of quantum gravitational electrodynamics suggested that the mass of the axion is related to the Planck mass, i.e. $m/m_{Pl} \approx e^{-1/2\alpha}$, giving, $m \approx 10^{-4}$ eV. This also gave a theoretical justification for the present small value of the cosmological constant, i.e. $\Lambda_0/\Lambda_{Pl} \approx e^{-2/\alpha} \approx 10^{-122}$.

Gravitational wave astronomy can provide an independent way to determine the Hubble's constant without depending on standard candles. The flux generated by gravitational waves is given by:

$$F_{GW} = \frac{c^3}{32\pi G}\omega^2 h^2 = \left(\frac{h}{10^{-22}}\right)^2 \left(\frac{\omega}{10^3 \, \text{Hz}}\right)^2 \times 3 \times 10^{-5} \, \text{W/m}^2 \tag{6.1}$$

i.e. $F_{GW} = 3 \times 10^{-5} \, \text{W/m}^2$, for $h = 10^{-22}$, $\omega = 1$ kHz.

The two observables are $h$ (the dimensionless strain $\frac{\delta l}{l}$) and $\omega$ (wave frequency). This gives $F_{GW}$. Total energy (generated) emitted by a source at a distance $D$ with received flux $F_{GW}$ is:

$$E = F_{GW} \times 4\pi D^2 \times t_d \tag{6.2}$$

$D$ is unknown, $t_d$ is the duration of GW burst or event. We have another relation connecting $h$ and $E$, i.e.

$$h \approx \frac{GE}{c^4 D} \approx \frac{G F_{GW} 4\pi D^2 t_d}{c^4 D} \approx \frac{G F_{GW} 4\pi D t_d}{c^4} \tag{6.3}$$

$h$ and $F_{GW}$ thus gives $D$. From Hubble's law, we have $v = H_0 D$.

In the merger of a neutron star binary, a gamma-ray burst is simultaneously released, with an afterglow. The redshift of the electromagnetic radiation (from this simultaneous occurrence) gives the velocity of the host galaxy. Thus for redshift $z$,

$$(1+z) = \frac{(c+v)^{1/2}}{(c-v)^{1/2}} \tag{6.4}$$

This gives $v$, since $z$ is measurable. So simultaneous observation of GW and electromagnetic radiation gives $H_0$ from the above relation. Thus, $D = \frac{8c}{h\omega^2 t_d}$, $H_0 = \frac{h\omega^2 t_d}{8c} v$, and this does not depend on standard candles like supernovae.

## 6.3  Constraints on Gravitational Dipole Radiation from Pulsars

The possibility of the existence of long range vector forces in addition to electromagnetism has been raised now and then, starting from Lee and Yang, who envisaged such a force coupling to baryon number. More recently, there have been several experiments based on various theoretical considerations to search for intermediate range vector forces coupling to isospin or hypercharge. These forces would violate the Einstein equivalence principle, and tight limits on their coupling relative to gravity are implied by such experiments. Deviations from Newton's inverse square law over short sub-millimetre distance scales are now being sought mainly motivated by ideas of large extra dimensions, forces due to dark matter particles like axions and Casimir type effects, due to dark energy, etc.

More recently, in an attempt to give a relativistic basis for MOND, which has had a good amount of phenomenological success in describing galaxy flat rotation curves without involving dark matter, a tensor-vector-scalar (TeVeS) theory has been proposed as an alternative to Einstein's general relativity. Cosmological tests for such modified theories have been proposed. These types of theories also involve a long range vector component of gravity. The laboratory tests of additional long range vector forces are essentially static experiments, mainly looking for a violation of the equivalence principle through an Eötvös type torsion balance or searching for deviations from Newtonian inverse square law.

However, such additional vector forces would have radiative effects in competition with gravitational radiation from astronomical sources, especially from binary systems consisting of compact objects. Indeed constraints on the strength of such forces were put from the known accuracy to which general relativity has been tested for the binary pulsar as regards the changing separation (changing period) between the components.

In analogy with electromagnetism, we would expect additional vector forces to give rise to electric and magnetic dipole radiation, which would cause additional effects in slowing down compact objects like neutron stars. As pulsar spins and periods have been determined to a high level of accuracy, this could be used to stringently constrain the strength of such forces. If $M$ be the total mass, $R$ the separation and $\omega$ the angular frequency of revolution, then the quadrupole energy loss due to gravitational radiation is,

$$\dot{E}_Q = \frac{32G}{5c^5} M^2 R^4 \omega^6 \tag{6.5}$$

Vector forces would, in addition, give rise to dipole radiation. If there is a gravitational dipole of relative strength $\alpha$ as compared to gravity ($\alpha \ll 1$) then the emission rate of gravitational dipole radiation is given by:

$$\dot{E}_D = \frac{2G}{3c^3} \alpha M^2 R^2 \omega^4 \tag{6.6}$$

So for a typical value of $M \approx 1.4\, M_\odot$, $M \approx 20$ km (typical binary pulsar system), and the fact that there is an agreement to 0.1% (with quadrupole radiation) implies a constraint on the strength, i.e. $\alpha$ of,

$$\alpha < \frac{v^2}{c^2} 10^{-3} < 10^{-10} \qquad (6.7)$$

The (orbital) velocity $v$ for the binary pulsar is $\sim 300$ km/s.

Similar constraints would result from the 2.4 h period binary pulsar. The very precise slowdown of the millisecond pulsar, i.e. given by $\dot{P}/P \approx 10^{-19}\,\mathrm{s}^{-1}$ implies,

$$\alpha \frac{2G}{3c^3} M^2 R^2 \omega^4 < \frac{B^2 R^6 \omega^4}{6c^3} \qquad (6.8)$$

$B$ is typically $10^8$ G, for millisecond pulsars. Thus the millisecond pulsar slowdown implies $\alpha < 10^{-20}$. I.e. any dipole (vector) component of the gravitational field is $10^{-20}$ of the Newtonian field strength. For a collapsing star radiating energy, similar limits can be put from,

$$\dot{E}_D = \frac{G}{c^3} \alpha P \dot{M} \qquad (6.9)$$

Again the merger time due to gravitational radiation is $\sim \frac{5R_0^4 c^5}{256 M^2 \mu G^3}$, where $\mu$ is the reduced mass. This gives a time scale of a billion years for the Hulse–Taylor binary pulsar and a few million years for the 2.4 h binary pulsar. On the other hand, for dipole radiation, the merger time is proportional to $R_0^3$ and not $R_0^4$. This could have testable implications for widely separated binaries (like black hole binaries, such as OJ 287).

Dipole radiation would cause a rate of change in the position of the centre of mass of the system. The 'magnetic' part of $\dot{E}_D$ (as given by Eq. (6.6)) would be smaller by a factor of $v/c$ (i.e. $< 10^{-3}$ for the binary pulsars but would, however, be 0.1 for the millisecond pulsar). The 'chirping' exponent in analysing the increasing frequency as the binaries spiral is different for dipole. For energy loss proportional to $r^{-k}$, the chirp frequency is $(t_0 - t)^{-3/2(k-1)}$. For dipole, $k = 3$.

It must be pointed out that lunar laser ranging (LLR) currently provides the best constraint on deviation from Newtonian force law as $10^{-10}$ times the strength of gravity (at $10^8$ m scale). This is comparable to the above limits. The millisecond pulsar gives a much tighter limit, ten orders more stringent. The above astronomical limits would tightly constrain the coupling of such terms in the gravitational action.

## 6.4   Gravitational Waves from Cosmic Strings

Cosmic strings are linear topological defects formed in the early universe when temperatures (energies) were extremely large ($10^{14}$ GeV), resulting from the breakdown of symmetry (usually associated with the asymmetric potential of a self-interacting scalar field). The tension $\eta$ in the string is the mass (energy) per unit length, characterising the energy at which the symmetry is broken. The cosmic string expands with the universe, reaching lengths of perhaps even kilo-parsecs.

They can lose energy by gravitational waves. The gravitational wave power emitted by a cosmic string is $P_{GW} = \beta G c \eta^2$, where $\beta \approx 1$ is a numerical constant. The 'lifetime' of the cosmic string due to the emission of gravitational waves is $L/\beta G c \eta$, where $L$ is the length of the cosmic string. For the usual symmetry breaking energy scales (associated with their formation) $\eta$ is generally, $10^{-6} \frac{c^2}{G}$. This gives a tension of about $10^{21}$ kg/m. The geometry associated with the cosmic string (i.e. deformation due to the string mass) is that of a conical defect, with the angle deficit (from $2\pi$) of $\sim 8\pi G \eta / c^2$, in the $d\phi$ coordinate. The metric is,

$$ds^2 = c^2 dt^2 - dr^2 - dz^2 - r^2 \left( d\phi - \frac{8\pi G \eta}{c^2} \right) \tag{6.10}$$

Cosmic strings can be observational signatures as giving rise to 'wide-angle' gravitational lenses. Their effect on the CMBR is to distort the temperature as $\frac{\Delta T}{T} \approx \frac{4\pi G \eta}{c^2} \approx 10^{-6} - 10^{-5}$ (Stebbins–Whitford effect). There are, at present severe observational constraints on the present presence of cosmic strings in the universe; their contribution to the density parameter is $\ll 1$. So at present, we are not in a stringy universe.

# Problems on Gravitational Waves

1. For a white dwarf binary (each with a mass of one solar mass) having a 5-min orbital period at a distance of 100 pc, estimate the strain for a detector on Earth. How does the orbital period change with time due to gravitational radiation?

2. A black hole binary has components of masses $10^8 M_\odot$ and $10^6 M_\odot$. For what separation between the components will gravitational radiation from the system lead to a merger of the black hole on a Hubble time scale ($10^{10}$ years)? What would be the present frequency of emission of the gravitational waves from such a system?

3. Recently a white dwarf binary (components ~ 1 solar mass each) with a 5 min period has been discovered. Estimate the time scale for the merger of the two components due to gravitational radiation. Suppose such a system is 100 pc away; by how much would it change the length of a 3 km arm of a laser interferometer detector?

4. How can one define stokes parameters for a plane gravitational wave? From the three stokes parameters, how would you calculate the fraction of circular polarization, linear polarization and direction of maximum linear polarization?

5. An elastic beam (rod) is used to detect gravitational waves at their lowest normal mode frequency, $\omega_0$. Can one use the rod to detect harmonics, i.e., $\omega_n = n\omega_0$? If the rod has the same damping time for all modes, how does the ratio of maximum amplitude squared of the displacement to the energy flux of the wave vary with $n$? That is, what is the sensitivity of the nth mode relative to the 0th?

6. If an asymmetric explosion releases an energy of $E \approx 10^{47}$ J, estimate the number of gravitons emitted. If the explosion took place 1 Mpc away, what is the strain on the detector?

7. A thin metal rod of density $\rho \approx 10^4 \, \text{kg/m}^3$ is spinning at a frequency of $10^3$ Hz around a symmetrical perpendicular axis. Compare the gravitational radiation power emitted with the electromagnetic radiation emission, which could arise from the slight excess of electrons pushed by centrifugal forces towards the ends of the rod.

© The Author(s), under exclusive license to Springer Nature Switzerland AG 2023
A. Kenath and C. Sivaram, *Physics of Gravitational Waves*,
SpringerBriefs in Physics,
https://doi.org/10.1007/978-3-031-30463-7

8.  The eleven minute binary 4U 1820–30 consists of a neutron star and a helium white dwarf (the tightest known LMXB). Calculate

    a.  The mutual orbital velocity.
    b.  Gravitational radiation energy loss rate.
    c.  Rate of shrinking of orbital radius.
    d.  Merging time due to gravitational wave emission.
    e.  Energy and duration of the burst of gravitational radiation when the merger occurs.

9.  Calculate the separation for the binary black holes system OJ 287 that would enable them to merge in a Hubble time. Comment on the result.

10. For two black holes of masses 66 and 85 solar masses, what should be their separation so that they merge in half the Hubble age? Consider orbital eccentricity of (a) 0.5 (b) 0.1.

11. The upper general relativistic limit to radiation power luminosity is $c^5/G \approx 2 \times 10^{52}$ W. Show that no matter how large the masses of the merging black holes are, the power released cannot exceed this value.

12. Hawking's area theorem estimates the maximum energy that can be radiated by merging of two black holes of masses 66 and 85 solar mass. If they are spinning in opposite directions and each has an angular momentum 0.5 of the maximum, what is the energy emitted?

13. Plaskett's Star is 1.5 kpc away. What would be the strain on a LIGO detector when such a binary black hole merges?

14. At the distance of the Virgo Cluster, estimate the flux of gravitational waves, gamma-rays, and neutrinos at detectors on Earth.

15. If two $10^6 \, M_\odot$ SMBHs are separated by $10^{14}$ m, , what is their orbital period (circular orbits)? If they merge and assume ten per cent of the mass-energy goes into gravitational waves, calculate (i) the flux in GW energy at a distance of 1 Gpc. (ii) The dimensionless strain in LISA. (iii) Repeat the above for $10^9 \, M_\odot$ SMBHs (after merger).

16. Find the Hubble constant for the parameters $z = 2$, $h = 10^{-22}$, $\omega = 1$ kHz, and $t_d = 10^{-2}$ s.

17. Given the orbital period of 1201 s, estimate the time scale for the merger of the white dwarf binary system, J2322+0509, due to the emission of gravitational waves.

18. Show that the power spectrum of gravitational waves from 'violent' events in the early universe at the TeV scale of energies can be expected to peak around a frequency of about $10^{-3}$ Hz at the present epoch.

19. Estimate the flux for gravitational waves with a frequency of 100 Hz and $h = 10^{-22}$? How does it compare with the flux of reflected sunlight from Jupiter on Earth?

20. Find the lifetime due to the emission of gravitational waves by a cosmic string one kpc long.

# Bibliography

Abadie, J., Abbott, B. P., Abbott, T. D., Abernathy, M. R., Benacquista, M., Creighton, T., Daveloza, H., Diaz, M. E., Grosso, R., Mohanty, S., & Mukherjee, S. (2011). A gravitational wave observatory operating beyond the quantum shot-noise limit: Squeezed light in application. *Nature Physics, 7*, 962.

Abbott, B., Abbott, R., Adhikari, R., Ageev, A., Agresti, J., Allen, B., Allen, J., Amin, R., Anderson, S. B., Anderson, W. G., & Araya, M. (2005). Upper limits on gravitational wave bursts in LIGO's second science run. *Physical Review D, 72*, 062001.

Abbott, B. P., Abbott, R., Abbott, T. D., Abernathy, M. R., Acernese, F., Ackley, K., Adams, C., Adams, T., Addesso, P., Adhikari, R. X., & Adya, V. B. (2016). Observation of gravitational waves from a binary black hole merger. *Physical Review Letters, 116*, 061102.

Abbott, B. P., Abbott, R., Abbott, T. D., Acernese, F., Ackley, K., Adams, C., Adams, T., Addesso, P., Adhikari, R. X., Adya, V. B., Affeldt, C., Afrough, M., Agarwal, B., Agathos, M., Agatsuma, K., Aggarwal, N., Aguiar, O. D., Aiello, L., Ain, A., …Woudt, P. A. (2017). Multi-messenger observations of a binary neutron star merger. *The Astrophysical Journal, 848*, L12.

Abbott, R., Abbott, T. D., Abraham, S., Acernese, F., Ackley, K., Adams, C., Adhikari, R. X., Adya, V. B., Affeldt, C., Agathos, M., & Agatsuma, K. (2020). GW190412: Observation of a binary-black-hole coalescence with asymmetric masses. *Physical Review D, 102*, 043015.

Alcock, C., Allsman, R. A., Alves, D. R., Axelrod, T. S., Becker, A. C., Bennett, D. P., Cook, K. H., Dalal, N., Drake, A. J., Freeman, K. C., & Geha, M. (2000). The MACHO project: Microlensing results from 5.7 years of large magellanic cloud observations. *The Astrophysical Journal, 542*, 281.

Amaro-Seoane, P., Aoudia, S., Babak, S., Binetruy, P., Berti, E., Bohe, A., Caprini, C., Colpi, M., Cornish, N. J., Danzmann, K., & Dufaux, J. F. (2012). Low-frequency gravitational-wave science with eLISA/NGO. *Classical and Quantum Gravity, 29*, 124016.

Arkani-Hamed, N., Dimopoulos, S., & Dvali, G. (1998). The hierarchy problem and new dimensions at a millimeter. *Physics Letters B, 429*, 263–272.

Arun, K., Gudennavar, S. B., Prasad, A., & Sivaram, C. (2018). Alternate models to dark energy. *Advances in Space Research, 61*, 567.

Arun, K., Gudennavar, S. B., Prasad, A., & Sivaram, C. (2019). Effects of dark matter in star formation. *Astrophysics and Space Science, 364*, 24.

Arun, K., Gudennavar, S. B., & Sivaram, C. (2017). Dark matter, dark energy, and alternate models: A review. *Advances in Space Research, 60*, 166.

Beaufils, Q., Sidorenkov, L. A., Lebegue, P., Venon, B., Holleville, D., Volodimer, L., Lours, M., Junca, J., Zou, X., Bertoldi, A., Prevedelli, M., Sabulsky, D. O., Bouyer, P., Landragin, A., Canuel, B., & Geiger, R. (2022). Cold-atom sources for the Matter-wave laser Interferometric Gravitation Antenna (MIGA). *Scientific Reports, 12*, 19000.

© The Author(s), under exclusive license to Springer Nature Switzerland AG 2023     53
A. Kenath and C. Sivaram, *Physics of Gravitational Waves*,
SpringerBriefs in Physics,
https://doi.org/10.1007/978-3-031-30463-7

Bekenstein, J. D. (2004). Relativistic gravitation theory for the modified Newtonian dynamics paradigm. *Physical Review D, 70*, 083509.

Bender, P., et al. (1994). *LISA, Laser interferometer space antenna for gravitational wave measurements: ESA assessment study report*. R. Reinhard, ESTEC.

Bertotti, B., & Sivaram, C. (1991). Radiation of the fifth-force field. *Il Nuovo Cimento B, 106*, 1299.

Biscoveanu, S., Talbot, C., Thrane, E., & Smith, R. (2020). Measuring the primordial gravitational-wave background in the presence of astrophysical foregrounds. *Physical Review Letters, 125*, 241101.

Blair, D. G. (1991). *The detection of gravitational waves*. Cambridge University Press.

Blakeslee, J. P. (2003). Discovery of two distant type Ia supernovae in the Hubble deep field-north with the advanced camera for surveys. *Astrophysical Journal, 589*, 693.

Boughn, S., & Rothman, T. (2006). Aspects of graviton detection: Graviton emission and absorption by atomic hydrogen. *Classical and Quantum Gravity, 23*, 5839.

Brown, W. R., Kilic, M., Bédard, A., Kosakowski, A., & Bergeron, P. (2020). A 1201 s orbital period detached binary: The first double helium core white dwarf LISA verification binary. *The Astrophysical Journal Letters, 892*, L35.

Chen, P. (1991). Gravitational beamstrahlung. *Modern Physics Letters A, 6*, 1069.

Chincarini, G., Fiore, F., Della Valle, M., Antonelli, A., Campana, S., Covino, S., Cusumano, G., Giommi, P., Malesani, D., Mirabel, F., & Moretti, A. (2006). Gamma-ray bursts: Learning about the birth of black holes and opening new frontiers for cosmology. *The Messenger, 123*, 54.

Ciufolini, I. (2007). Dragging of inertial frames. *Nature, 449*, 41.

Clea, S., Arun, K., Sivaram, C., & Gudennavar, S. B. (2020). Effects of dark matter in red giants. *Physics of the Dark Universe, 30*, 100727.

Corda, C. (2009). Interferometric detection of gravitational waves: The definitive test for general relativity. *International Journal of Modern Physics D, 18*, 2275.

Dabrowski, Y., Fabian, A. C., Iwasawa, K., Lasenby, A. N., & Reynolds, C. S. (1997). The profile and equivalent width of the X-ray iron emission line from a disc around a Kerr black hole. *Monthly Notices of the Royal Astronomical Society, 288*, L11.

De Logi, W. K., & Mickelson, A. R. (1977). Electrogravitational conversion cross sections in static electromagnetic fields. *Physical Review D, 16*, 2915.

Fischbach, E., Sudarsky, D., Szafer, A., Talmadge, C., & Aronson, S. H. (1986). Reanalysis of the Eötvös experiment. *Physical Review Letters, 56*, 1427.

Flowers, E., & Itoh, N. (1979). Transport properties of dense matter. II. *The Astrophysical Journal, 230*, 847.

Gelmini, G. B. (2006). DAMA detection claim is still compatible with all other DM searches. *Journal of Physics: Conference Series, 39*, 040.

Giazotto, A. (1989). Interferometric detection of gravitational waves. *Physics Reports, 182*, 365.

Grindlay, J., Zwart, S. P., & McMillan, S. (2006). Short gamma-ray bursts from binary neutron star mergers in globular clusters. *Nature Physics, 2*, 116.

Häkkinen, H., Moseler, M., Kostko, O., Morgner, N., Hoffmann, M. A., & Issendorff, B. V. (2004). Symmetry and electronic structure of noble-metal nanoparticles and the role of relativity. *Physical Review Letters, 93*, 093401.

Hoyle, C. D., Schmidt, U., Heckel, B. R., Adelberger, E. G., Gundlach, J. H., Kapner, D. J., & Swanson, H. E. (2001). Submillimeter test of the gravitational inverse-square law: A search for "large" extra dimensions. *Physical Review Letters, 86*, 1418.

Huang, X. J., Zhang, W. H., & Zhou, Y. F. (2016). 750 GeV diphoton excess and a dark matter messenger at the Galactic Center. *Physical Review D, 93*, 115006.

Laine, S., Dey, L., Valtonen, M., Gopakumar, A., Zola, S., Komossa, S., Kidger, M., Pihajoki, P., Gómez, J. L., Caton, D., & Ciprini, S. (2020). Spitzer observations of the predicted Eddington flare from Blazar OJ 287. *The Astrophysical Journal Letters, 894*, L1.

Lee, T. D., & Yang, C. N. (1955). Conservation of heavy particles and generalized gauge transformations. *Physical Review, 98*, 1501.

Long, J. C., Chan, H. W., & Price, J. C. (1999). Experimental status of gravitational-strength forces in the sub-centimeter regime. *Nuclear Physics B, 539*, 23.

Longair, M. S. (2010). *High energy astrophysics*. Cambridge University Press.

Lyne, A. G., Burgay, M., Kramer, M., Possenti, A., Manchester, R. N., Camilo, F., McLaughlin, M. A., Lorimer, D. R., D'Amico, N., Joshi, B. C., & Reynolds, J. (2004). A double-pulsar system: A rare laboratory for relativistic gravity and plasma physics. *Science, 303*, 1153.

Lyne, A. G., & Smith, F. G. (1992). *Pulsar astronomy*. Cambridge University Press.

Milgrom, M. (1983). A modification of the Newtonian dynamics as a possible alternative to the hidden mass hypothesis. *The Astrophysical Journal, 270*, 365.

Montero-Camacho, P., Fang, X., Vasquez, G., Silva, M., & Hirata, C. M. (2019). Revisiting constraints on asteroid-mass primordial black holes as dark matter candidates. *Journal of Cosmology and Astroparticle Physics, 2019*, 031.

Narain, G., Schaffner-Bielich, J., & Mishustin, I. N. (2006). Compact stars made of fermionic dark matter. *Physical Review D, 74*, 063003.

Novikov, I. D., Kardashev, N. S., Shatskii, A. A., Lukash, V. N., & Mikheeva, E. V. (2007). Scientific session of the Physical Sciences Division of the Russian Academy of Sciences. *Physics-Uspekhi, 50*, 965.

Page, D. N., & Hawking, S. W. (1976). Gamma-rays from primordial black holes. *The Astrophysical Journal, 206*, 1.

Peacock, J. A. (1999). *Cosmological physics*. Cambridge University Press.

Piran, T. (2005). The physics of gamma-ray bursts. *Reviews of Modern Physics, 76*, 1143.

Raffelt, G., & Stodolsky, L. (1988). Mixing of the photon with low-mass particles. *Physical Review D, 37*, 1237.

Sartore, N., Ripamonti, E., Treves, A., & Turolla, R. (2010). Galactic neutron stars-I. Space and velocity distributions in the disk and in the halo. *Astronomy & Astrophysics, 510*, A23.

Shapiro, S. L., & Teukolsky, S. A. (1983). *Black holes, white dwarfs, and neutron stars: The physics of compact objects*. Wiley.

Sivaram, C. (1984). Thermal gravitational radiation from stellar objects and its possible detection. *Bulletin of the Astronomical Society of India, 12*, 350.

Sivaram, C. (1990). Plasma damping of gravitational waves. In E.R. Priest & V. Krishnan (Eds.), *IAU Symposium, No. 142* (p. 62). Kluwer Academic Publishers.

Sivaram, C. (1993). On the Maxwellian alternative to the galactic dark matter problem. *Astronomy and Astrophysics, 275*, 37.

Sivaram, C. (1999). Constraints on the photon mass and charge and test of equivalence principle from GRB 990123. *Bulletin of the Astronomical Society of India, 27*, 627.

Sivaram, C. (2000). Some implications of quantum gravity and string theory for everyday physics. *Current Science, 79*, 413.

Sivaram, C. (2001). Black hole hawking radiation may never be observed! *General Relativity and Gravitation, 33*, 175.

Sivaram, C. (2007). Dark energy may link the numbers of Rees. arXiv preprint arXiv:0710.4993

Sivaram, C. (2008). OJ 287: New testing ground for general relativity and beyond. arXiv preprint arXiv:0803.2077

Sivaram, C. (2015). Gravitational waves: Some less discussed intriguing issues. *International Journal of Modern Physics D, 24*(12), 1544023.

Sivaram, C., & Arun, K. (2009a). New class of dark matter objects and their detection. arXiv preprint arXiv:0910.2306

Sivaram, C., & Arun, K. (2009b). Relativistic jets from black holes: A unified picture. arXiv preprint arXiv:0903.4005

Sivaram, C., & Arun, K. (2010). Charged black holes and constraints on baryon asymmetry. arXiv preprint arXiv:1003.1667

Sivaram, C., & Arun, K. (2011a). New class of dark matter objects and their detection. *Open Astronomy Journal, 4*, 57.

Sivaram, C., & Arun, K. (2011b). Thermal gravitational waves from primordial black holes. *The Open Astronomy Journal, 4*, 72.

Sivaram, C., & Arun, K. (2011c). Thermal gravitational waves. *The Open Astronomy Journal, 4*, 65.

Sivaram, C., & Arun, K. (2014). Intermediate mass black holes: Their motion and associated energetics. *Advances in High Energy Physics, 2014*, 924848.

Sivaram, C., & Arun, K. (2019). Dark matter objects: Possible new source of gravitational waves. *Earth, Moon, and Planets, 123*, 9.

Sivaram, C., Arun, K., & Kiren, O. V. (2016). Planet Nine, dark matter and MOND. *Astrophysics and Space Science, 361*, 1.

Sivaram, C., Arun, K., & Kiren, O. V. (2018). Forming supermassive black holes like J1342+0928 (invoking dark matter) in early universe. *Astrophysics and Space Science, 363*, 1.

Sivaram, C., Arun, K., & Samartha, C. A. (2008). Phase-space constraints on neutrino luminosities. *Modern Physics Letters A, 23*, 1470.

Spergel, D. N., Verde, L., Peiris, H. V., Komatsu, E., Nolta, M. R., Bennett, C. L., Halpern, M., Hinshaw, G., Jarosik, N., Kogut, A., & Limon, M. (2003). First-year Wilkinson Microwave Anisotropy Probe (WMAP)* observations: Determination of cosmological parameters. *The Astrophysical Journal Supplement Series, 148*, 175.

Stacey, F. D., Tuck, G. J., Moore, G. I., Holding, S. C., Goodwin, B. D., & Zhou, R. (1987). Geophysics and the law of gravity. *Reviews of Modern Physics, 59*, 157.

Taylor, J. H., Wolszzan, A., Damour, T., & Weisberg, J. M. (1992). Experimental constraints on strong-field relativistic gravity. *Nature, 355*, 132.

Thorne, K. S., Misner, C. W., & Wheeler, J. A. (2000). *Gravitation*. Freeman.

Tisserand, P., Le Guillou, L., Afonso, C., Albert, J. N., Andersen, J., Ansari, R., Aubourg, É., Bareyre, P., Beaulieu, J. P., Charlot, X., & Coutures, C. (2007). Limits on the macho content of the galactic halo from the EROS-2 survey of the magellanic clouds. *Astronomy & Astrophysics, 469*, 387.

Tokuoka, T. (1975). Interaction of electromagnetic and gravitational waves in the weak and short wave limit. *Progress of Theoretical Physics, 54*, 1309.

van Leeuwen, F. (Ed.). (2007). *Hipparcos, the new reduction of the raw data*. Springer.

Verde, L., Peiris, H. V., Spergel, D. N., Nolta, M. R., Bennett, C. L., Halpern, M., Hinshaw, G., Jarosik, N., Kogut, A., Limon, M., & Meyer, S. S. (2003). First-year Wilkinson Microwave Anisotropy Probe (WMAP)* observations: Parameter estimation methodology. *The Astrophysical Journal Supplement Series, 148*, 195.

Weinberg, S. (1972). *Gravitation and cosmology*. Wiley.

Weisberg, J. M., & Taylor, J. H. (2002). General relativistic geodetic spin precession in binary pulsar B1913+16: Mapping the emission beam in two dimensions. *The Astrophysical Journal, 576*, 942.

Will, C. M. (2014). The confrontation between general relativity and experiment. *Living Reviews in Relativity, 17*, 1.

Williams, J. G., Turyshev, S. G., & Boggs, D. H. (2004). Progress in lunar laser ranging tests of relativistic gravity. *Physical Review Letters, 93*, 261101.

Zel'dovich, Y. B. (1973). Primordial magnetic fields and gravitational wave effects on CMBR. *Soviet Physics JETP, 38*, 652.

Zeldovich, Y. B., & Novikiv, I. (1983). *Relativistic astrophysics*. University of Chicago Press.

Zhao, H., Bacon, D. J., Taylor, A. N., & Horne, K. (2006). Testing Bekenstein's relativistic modified Newtonian dynamics with lensing data. *Monthly Notices of the Royal Astronomical Society, 368*, 171.